RESEARCH WORKS AND DISCOVERIES BY:
EZZAT E. MAJD POUR, M.D.

CHEMISTRY OF FUNDAMENTAL PARTICLES PHYSICS OF FUNDAMENTAL PARTICLES

NUCLEAR AND MEDICAL SCIENCES OF FUNDAMENTAL PARTICLES

outskirts
press

CHEMISTRY OF FUNDAMENTAL PARTICLES PHYSICS OF FUNDAMENTAL PARTICLES
Nuclear and Medical Sciences of Fundamental Particles
All Rights Reserved.
Copyright © 2020 Research Works and Discoveries by: Ezzat E. Majd Pour, M.D.
v1.0

The opinions expressed in this manuscript are solely the opinions of the author and do not represent the opinions or thoughts of the publisher. The author has represented and warranted full ownership and/or legal right to publish all the materials in this book.

This book may not be reproduced, transmitted, or stored in whole or in part by any means, including graphic, electronic, or mechanical without the express written consent of the publisher except in the case of brief quotations embodied in critical articles and reviews.

Outskirts Press, Inc.
http://www.outskirtspress.com

ISBN: 978-1-9772-1864-3

Cover Photo © 2020 Ezzat E. Majd Pour, M.D.. All rights reserved - used with permission.

Outskirts Press and the "OP" logo are trademarks belonging to Outskirts Press, Inc.

PRINTED IN THE UNITED STATES OF AMERICA

THE FIRST VOLUME:

This book is the results of the personal research works and new discoveries of the Author. Which no - body has discovered these findings during the past world history, except the Author who has done those, during the last several decades through continuous research works.

The Author through seventy - years research works days and nights, discovered the Electron Nucleon fundamental particle constructions, the Particle cloud –genesis, the fundamental particle's Chemistry, Physics, Biology, Psyche-genesis, Thought - current –genesis, cause of the psychiatric diseases, and many other fundamental Particle transmitted diseases, Plant's central and peripheral intelligence system centers, particle weapons of mass destructions, etc.

The Author's discovery rights all are protected and reserved through the United States copyright office for him, and owner of these creations discoveries is Author only. no parts of these personal research works, discoveries, and properties of author allowed for printing, reproductions, distributions without written permission from Author.

This book must not be reproduced, printed, stored, retrieved, copied, recorded, transmitted digitally or by any other means and the ways, Without written permission from the author personally. Teaching these discoveries through this book in classrooms are exceptions.

EZZAT E. MAJD POUR, M.D.

INTRODUCTION

Through last seventy – years research works, I discovered Mental disorders produced through transmission of Particle Clouds from sick individuals to normal ones, and psychiatric disorders are Particle Cloud (S. Y. T. E.– F.P. – I.I. – P. cl.) transmitted diseases. the sick Patients emit abnormal Sonic- Light particle clouds to air, abnormal Particle Clouds travel through air, inter into recipients normal CNS internal Electron Nucleons, and infect the recipient's internal Electron Nucleon particle compounds. and cause electron Nucleons infections in the recipients. Psychosis transmit psychotic pattern Particle Clouds, Depressive transmit depressive compound Particle Clouds, and so on.

During Life time research I discovered, the different animal species emit Particle Clouds (S.Y. T.E. – F.P. – I.I. - P. cl.) from their CNS Electrons Nucleons to the outside world air, these emitted Particle Clouds through air travel toward recipient individuals through Ex. – PCS circuits, Particle Clouds inter into recipient's CNS Electrons Nucleon, and combine with their particle –compounds, construct recipient's Electrons Nucleons particle compound – constructions.

Above is the basics of the Psyche – genesis, Thought - current –genesis, learning knowledge and science, social, religious, and educational material accumulations inside the CNS Electrons Nucleons. Through this phenomenon the individuals in home, school, college, university classes, etc. communicate and accumulate information and images Particle Clouds of the outside world, inside their CNS Electrons Nucleons, these Particle Cloud storages and retrievals inside CNS Electrons Nucleons, is intelligence systems –genesis phenomenon, and used in daily communications between different individuals, different animal, circulate air-born in Ex. PCS – circuits.

Through the research works, I discovered F.P. general chemistry, F. P. organic chemistry, F.P. biological chemistry, A. S. I. F.P. Mol. C.I.C., the F. P. biological sciences, Particle – Colonies, and Fundamental Particle Evolution paths and orders which in continuation constructed the living things.

I discovered the Unit of Graviton, the Unit of Mass, the Unit of Light Energy, the Unit of Electric Energy, the Unit of Thermal Energy, etc. also I discovered chemical compound constructions of the Electrons, Nucleons and other Nano- Units.

Also I discovered the Particle Circulation Systems(PCS) of Nucleus, Peripheron, and Atom. The PCS inside Electrons, the PCS inside the Nucleons, and PCS in P.A.S.

I discovered Light fundamental particle constructed Atoms, the Electric F.P. constructed Atoms, the Thermal F.P.

constructed Atoms, etc. chemical compound constructions of Nano – Units, and Quantum energy provider Fundamental Particles for Nano- Sites.

During the seven Decades research works, I discovered Plant's Central Intelligence System centers and Peripheral Intelligence System centers in plants. Also I discovered use of bio-hostile fundamental Particle weapons of Mass destructions in international wars and otherwise, all destroy and damage internal Electron Nucleons particle compound constructions of the CNS Electrons Nucleons, the use of these bio-hostile microwave – laser Weapons of Mass destructions are crimes against Humanity, I appeal for their removal all over the world, and their use is illegal even in international wars.

I discovered that the Atom's Turn-over process is equal to the Atom – genesis, during that Atom's molecular Evolution era orders and paths. Also I discovered the Fetus – Genesis during Embryo, is equal to the Animal Plant – genesis, which occurred under the molecular Evolution orders and paths, during the world's past Evolution History.

PHENOMENON OF ANIMAL-GENESIS AND PLANT –GENESIS IN EMBRYO

THROUGH THE LONG PERIODS OF THE CONTINUOUS RESEARCH WORKS, I DISCOVERED IN ANIMAL'S AND PLANT'S EMBRYO, THE WELL DEFINED FIX CHEMICAL FORMULARY CHANGES OF MEDIA AROUND THE STEM CELLS OF THE FETUS, PRODUCE THE STABLE - STEM CELLS - MUTATION IN EMBRYO. THIS PHENOMENON IS TRUE FOR BOTH ANIMAL AND PLANTS EARLY STAGE FETAL GROWTH RAPID CELL DIFFERENTIATIONS.

ADDITIONALLY, EACH GIVEN WELL DEFINED NEW FIXED STEM CELL'S MUTATIONS IN FETAL STEM CELLS DURING EMBRYO, SECONDARILY PRODUCE THE STEM CELL'S MEDIA CHEMICAL FORMULARY TO CHANGE INTO ANOTHER KIND WELL DEFINED FIXED AND STABLE NEW MEDIA CHEMICAL FORMULARIES. AND THESE CYCLES OF THE ONE CHANGE TRIGGERS THE OTHER CHANGE CONTINUES IN CLOSED CYCLES ONE AFTER THE OTHER AND CAUSE STEM CELL'S DIFFERENTIATIONS, WHICH IS THE CAUSE OF GENESIS OF DIFFERENT TISSUES FOR FETUS IN EMBRYO.

THIS IS PHENOMENON OF FETAL ALTERNATING STABLE MUTATIONS, WHEN THEY ARE FOLLOWED BY PRODUCTIONS OF THE NEW STABLE FIXED MEDIA CHEMICAL FORMULARY CHANGES IN EMBRYO. THE FIRST MUTATION CYCLICALLY ALTERNATE AND PRODUCE THE SECOND ONES CHEMICAL FORMULARY CHANGES, AND VICE VERSA, UNDER CLOSED CYCLES ONE AFTER THE OTHER. AND PRODUCE NEW OTHER KIND STEM CELL – GENESIS, AND TISSUE DIFFERENTIATIONS WHICH ARE NEEDED TO CONSTRUCT DIFFERENT MULTI - TISSUE CONSTRUCTION: FETAL ORGANS AND SYSTEMS UNDER THE MOLECULAR EVOLUTION ORDERS AND PATHS.

IN FETUS, THE MEDIA CHEMICAL FORMULARY CHANGE CAUSE THE STEM CELL MUTATIONS, AND THE STEM CELL MUTATIONS CAUSE THE PRODUCED MUTANTS MEDIA CHEMICAL FORMULARY CHANGE INTO ANOTHER KIND IN A WELL DEFINED FIX STRUCTURE CLOSED CYCLES ALTERNATIONS. WHICH THESE MUTATIONS TAKE PLACE IN EMBRYO AUTONOMOUSLY, SEQUENTIALL, WITH EXPONENTIAL SPEEDS OF THE STABLE NEW MUTATION PRODUCTIONS AND CELL DIFFERENTIATIONS, EXPONENTIAL STEM CELLS DUPLICATIONS, NEW TISSUE PRODUCTIONS, WHICH IS THE CAUSE FOR: BODY ORGANS –GENESIS, NEW BODY SYSTEMS – GENESIS, AND TOTAL NEW ANIMAL –GENESIS. AND PLANT –GENESIS.

THE IMPORTANCE OF THESE PHENOMENONS ARE IN THE FUTURE IN CHEMICAL LAB.S, THESE PATHS MAY BE USED ONE DAY, TO CULTURE ANIMAL ORGANS -GENESIS AND ANIMAL SYSTEMS - GNESIS IN LABORATORIES FOR HUMAN

TOTAL SYSTEM OR ORGAN EXCHANGES THROUGH THE SURGICAL PROCEDURES, AND EXTEND THE HUMAN LIVES FOR SEVERAL ADDITIONAL HUNDRED YEARS.

EZZAT EL MAJD POUR, M.D.

Contents

Introduction:

The Creations and discoveries have been presented at this book, that nobody of the past, had capabilities to imagine these discoveries, and creations, since The inceptions of the Earth.

Ezzat E. Majd Pour, M. D.
My email address: ezzatmajd@icloud.com

	Pages:
Fundamental particles	1
Fundamental particle's colonies.	2
Earth's Fundamental Particles (FP)	3
Sensible Fundamental Particles	4
Classification of Fundamental Particles	5 - 10

Light Fundamental Particles, Sonic Particles,
Electric FP, Thermal Fundamental Particles, etc.
Quantum Energy- providers into Nano Sites,
Energy Sources to Nano - locations,
Multi Energy- FP, T. FP, Y. FP, E. FP, S. FP,
Energy Providers into Elections, Nucleons,
Atoms, Quantum- Sites. ....11 - 15

Energy Storage inside Elections, Nucleons,
Energy release from Electrons, Nucleons,
Energy Emitting Nano Units, Energy Providers
to Nano Locations. ....15 - 21

Energy provider FP. (Y. -S. -T.- E.- Particles, and different
kinds Nano -Energy sources),
Units of light Energy, Units of Electric Energy,
Units of Thermal Energy, Units of Sonic Energy, unit of
Graviton, etc. 22-24

How FP operates body organs, and body systems of living
things, functions of body organ's Electrons, and Nucleons,
and Body Organ's CIS, and body Organ's PCS. ....25 - 30

Nano Neurology, Hierarchical FP intelligence systems, (PIS)
centers, and Particle Circulation Systems, (PCS). And Indigenous particle circulation systems I- PCS, ...31 - 40

Psyche Genesis, thought currents Genesis,................ 41-44

Universe Under Laws and Orders of Graviton,
the Units of the Graviton. 45- 49

Units of the Mass 50

Energy Units, Units of different kind Energies 51

Fundamental particle Clouds (P Cl.), ... 51-59
FP - I. I.- P. Cl. Genesis, P Cl. Storages inside the Elections and Nucleons, P Cl. Retrieval from inside Electrons and Nucleons to the outside Atoms, Electron Genesis, and Nucleon Genesis.

Thought Currents Genesis, and Psyche Genesis, ... 60 - 62

Internal Nucleus PCS, inter - Peripheron PCS, Peripheral Atom Space - PCS., ... 63 - 67

Nano Units, Electron Genesis, Nucleon's Genesis, and Genesis of Nano Units. .... 68 - 71

Y. FP. Constructed Electrons, Nucleons, Atoms, E. FP construction Electrons, Nucleons, Atoms, Thermal FP constructed : e-, P+, N•, Atoms. 72-77

Atom's Turn Over Phenomenon. 78- 79

Molecular Evolution, and creation of the life. 80-82

The introduction for the General Chemistry of Fundamental Particles 83-86

The Electron's Cycles 86-88

The Introduction to the Organic Chemistry of the Fundamental Particles. 89-90

The introduction to the Fundamental Particles Biological Chemistry, the Storages and Retrieving of the S. Y. E. T. - FP- I. I.- P. Cl. From CIS Electrons, Nucleon's 89-95

Introduction to the Nano Unit's Chemistry:
the Proton's Neo- Genesis,
the Neutron's Neo- Genesis,
the Electrons Neo Genesis.
The Positron's Neo Genesis, Neutrinos and anti Neutrino Neo Genesis 96-102

The Fundamental Particles Biological sciences in Plants, ... 103-108

The Alternating Stem Cells Mutations, with the Cell's Media chemical Formulary changes are the causes of Neo - Embryonic Genesis in living things, and Plants. 109-121

The Particle Clouds Genesis, The communications of different Electrons, Nucleons with each other. Storages of S. Y. E. T. -FP- I. I. - P. Cl. And storages of knowledge , and sciences inside The CNS Electrons, Nucleons, and retrieving of particles clouds to the outside CNS Electrons and Nucleons, or the memorization phenomenon of CNS. The psyche Genesis. And thoughts Genesis, which are particle clouds interactions. And currents in CNS. 121-140

S. Y. T. E.- FP- I. I. - P. Cl. Particle Clouds Circulation systems, or thought circulation Systems , Psyche Genesis. ... 141-143

Abnormal Y. S. T. E. -FP- I.I. - P. Cl. Transmission disorders, or Psychiatric diseases. Mental disorders are Particle transmitted diseases. 144-149

The CNS. FP- I. I. - P. Cl. Circulation systems, the A. S. I. FP. Mol. C. I. C. Between The CNS different Electron's Nucleon's FP.-I. I. - P. Cl. Etc. Produce thoughts currents and Psyche, ... 149-159

Storages and retrieval of the S. Y. - FP - I. I. - P. Cl. Inside the Silicon's Electron's Nucleons under A. S. I. FP. Mol. C.I. C. Through computer technologies. (The Intelligence atoms), and (non- intellectual atoms). 159-161

Intelligence Systems Centers of Plants 162-167

Plant's biological, physical, Chemical Functions takes place under the Plant's particle Intelligence Systems Centers (PIS). 168-173

The Plant's Immunity and
Plant's Defensive systems. 174-180

Plant's Central Particles Intelligences systems centers, (PIS). Hierarchy systems between the plant's PIS different systems centers. 180-182

Microwave-Laser Weapons of Mass Destruction's use in Wars, also against ordinary Citizens, inside their homes and Residences when they are in sleep by rulers , injuries, disabilities, murders and crimes against Humanity, War Syndromes, therapy, prevention, request for banning of these weapons of Mass Destruction. 183-203

Other Summaries from the Other discoveries of the Author 204-206

THE FUNDAMENTAL PARTICLES (F.P.)

THE EXISTING LAWS AND ORDERS OF MATTER IN UNIVERSE

DEFINITION AND INTRODUCTION:

FUNDAMENTAL PARTICLES ARE UNITS OF THE MATTER, THE FUNDAMENTAL PARTICLES CONSTRUCT EXISTING DIFFERENT ELECTRONS, NUCLEONS PARTICLE COMPOUND CONSTRUCTIONS OF EXISTING THINGS IN ALL OVER UNIVERSE. EACH GIVEN FUNDAMENTAL PARTICLE IN ANY GIVEN PLANET POSSESSES WELL- DEFINED, PARTICLE - SPECIFIC RELATIVELY FIXED PHYSICAL, CHEMICAL, AND BIOLOGICAL PROPERTIES, THE FUNDAMENTAL PARTICLES OF DIFFERENT PLANETS ARE DIFFERENT FROM EACH OTHER, BUT EACH GIVEN PLANET POSSESSES PLANET- SPECIFIC, NUMEROUS DIFFERENT FUNDAMENTAL PARTICLES CLASSES, INDIGENOUS FOR THAT PLANET AT MOST.

THE FUNDAMENTAL PARTICLE'S WEIGHT, GRAVITY, FREQUENCY, WAVE-LENGTH, POLE, SPEED, SPIN, TRANSIT ROUTES, TILTS, ROTATIONS, COLOR, SIZE SHAPE, PARTICLE-COLONY SHAPE, PARTICLE COLONY SIZE, SENSIBILITY,

PARTICLE CURRENTS, ETC., ARE ALL FUNDAMENTAL PARTICLE – SPECIFIC.

THE FUNDAMENTAL PARTICLES OF DIFFERENT PLANETS ARE DIFFERENT FROM EACH OTHER, THE FUNDAMENTAL PARTICLES AT DIFFERENT PLANETS CONSTRUCT DIFFERENT KIND PARTICLE COMPOUNDS, DIFFERENT ELECTRONS, NUCLEONS, AND ATOMS, UNDER AUTONOMOUS SEQUENTIAL INTER FUNDAMENTAL PARTICLE MOLECULAR CHEMICAL INTERACTION CYCLES AND CHAINS (A. S. I. F.P. Mol. C.I.C.). SOME PLANETS FUNDAMENTAL PARTICLES ARE BIOFRIENDLY AND SENSIBLE, MANY OTHERS PARTICLES ARE NOT SENSIBLE, MOST OTHER PLANETS PARTICLES ARE BIOHOSTILE, AND MANY ARE BIO-LETHAL PARTICLES.

FUNDAMENTAL PARTICLE COLONY.

PARTICLE POPULATION OF A COLONY.

FUNDAMENTAL PARTICLES TENDS TO COLONIZE IN WAVE – SHAPE FORMS, F.P. TRAVEL IN WAVE –SHAPE FORMS.

ONE FUNDAMENTAL PARTICLE COLONY EQUALS TO TOTAL FUNDAMENTAL PARTICLES POPULATION NUMBER IN ONE WAVE LENGTH FUNDAMENTAL PARTICLE, IN OTHER WORDS TOTAL PARTICLE NUMBERS THAT EXIST IN ONE WAVE – LENGTH PARTICLE IS EQUAL TO ONE COLONY FUNDAMENTAL-PARTICLE.

DIFFERENT FUNDAMENTAL PARTICLES HAVE DIFFERENT

PARTICLE POPULATION NUMBERS PER COLONY AND DIFFERENT PARTICLE COLONIES, DIFFERENT WAVE –LENGTH SIZES SIMILAR TO DIFFERENT COLONY SIZES OF DIFFERENT KIND FUNDAMENTAL PARTICLES ARE DIFFERENT FROM EACH OTHER.

THE FUNDAMENTAL PARTICLE COLONY SIZE AND TOTAL FUNDAMENTAL PARTICLE-POPULATIONS NUMBERS AT EACH WAVE –LENGTH OF A GIVEN FUNDAMENTAL PARTICLE CLASS ALWAYS ARE WELL DEFINED, FIXED AND PARTICLE SPECIFIC. AGAIN COLONIES OF DIFFERENT PARTICLES ARE DIFFERENT FROM EACH OTHER.

EARTH'S FUNDAMENTAL PARTICLES.

PLANET EARTHS FUNDAMENTAL PARTICLES ARE LIGHT WEIGHT (LESS F.P. – MASS), WEAK, BIOFRIENDLY, LESS ENERGY CONTENT, LESS GRAVITON FORCE, SOME ARE FLOATING AIRBORN PARTICLES, SOME F.P. ARE SENSIBLE, BUT MOST EARTH'S FUNDAMENTAL PARTICLES ARE NON SENSIBLE AND ARE NOT DISCOVERED YET, ALMOST MAJORITIES OF EARTHS PARTICLES ARE BIOFRIENDLY, WITH EXCEPTIONAL BIOHOSTILE PARTICLE AS WELL, THESE ARE THE F.P. CONSTRUCT EARTHS ELECTRONS NUCLEONS PARTICLE COMPOUND CONSTRUCTIONS FOR BOTH LIVE AND NON-LIVE ELECTRONS NUCLEONS AND ATOMS.

FUNDAMENTAL PARTICLES WEIGHTS, GRAVITON, ENERGY CONTENTS ARE PARTICLES – SPECIFIC, FOR EXAMPLE THERMAL PARTICLES (T.-F.P.), LIGHT PARTICLES (Y.- F.P.), SONIC

PARTICLES (S. –F.P.), ELECTRIC PARTICLES (E.-F.P.), X.-F.P., GAMMA FUNDAMENTAL PARTICLES ENERGIES, WEIGHTS, GRAVITON QUANTITIES AT EARTH ARE DIFFERENT FROM EACH OTHER, THESE PARTICLES IN COMPARISON TO BLACKS THERE IS THOUSANDS TO MILLIONS FOLDS DIFFERENCES, THOSE PARTICLES CONSTRUCT BLACK MATTER ATOMS FROM THEIR PARTICLE COMPOUNDS.

THE EARTH'S BIOHOSTILE FUNDAMENTAL PARTICLES SUCH AS GAMMA-PARTICLES, X- FUNDAMENTAL PARTICLES, INDUSTRIAL HIGH - ENERGY ELECTRIC FUNDAMENTAL PARTICLES, AND INDUSTRIAL HIGH ENERGY THERMAL - PARTICLES, ETC. AT EARTH MOSTLY ARE STORED INSIDE THE ELECTRONS AND NUCLEONS, IN PARTICLE COMPOUND CO-STRUCTIONS FORMS.

PLANET EARTH'S BIOFRIENDLY SENSIBLE PARTICLES
INTRODUCTIONS TO VISIBLE BIOFRIENDLY LIGHT
FUNDAMENTAL PARTICLES
AUDIBLE SONIC FUNDAMENTAL PARTICLES,
AND THERMAL PARTICLES,

THERE ARE NUMEROUS DIFFERENT KIND FUNDAMENTAL PARTICLE CLASSIFICATIONS IN PLANET EARTHS, FOLLOWING ARE SOME EXAMPLES:

1 - CLASIFICATIONS OF LIGHT PARTICLES OVER SENSIBILITIES AND NON-SENSIBILITIES CHARACTERS.

2 – CLASSIFICATIONS OF VISIBLE LIGHT FUNDAMENTAL PARTICLES OVER EXISTING DIFFERENCES IN LIGHT FUNDAMENTAL PARTICLES COLORS, SUCH AS: RED COLOR FUNDAMENTAL PARTICLES CLASSES, YELLOW COLOR LIGHT FUNDAMENTAL PARTICLES CLASS, GREEN COLOR LIGHT PARTICLES CLASS, BLUE COLOR FUNDAMENTAL PARTICLES CLASS, VIOLET COLOR FUNDAMENTAL PARTICLES, AND WHITE COLOR LIGHT PARTICLES, ETC.

ABOVE DIFFERENT CLASSES PARTICLE WAVE LENGTH, PARTICLE FREQUENCY, PARTICLE ENERGY KIND AND ENERGY QUANTITY, PARTICLES GRAVITON FORCE, PARTICLE COLONY SIZES, PARTICLE COLONY SHAPES, ETC., ALL ARE DIFFERENT IN DIFFERENT COLOR PARTICLES FROM EACH OTHER.

3 - LIGHT PARTICLES CLASSIFICATIONS OVER BEING BIOHOSTILE OR BIOFRIENDLY IN EARTH.

4- SONIC FUNDAMENTAL PARTICLES CLASIFICATION, EITHER ARE HUMAN SENSIBLE OR NOT.

5- CLASSIFICATION OF SONIC FUNDAMENTAL PARTICLES OVER BEING BIOFRIENDLY OR BIOHOSTILE.

6- PLANET EARTHS THERMAL FUNDAMENTAL PARTICLE CLASSIFICATIONS.

7- PLANET EARTHS ELECTRIC FUNDAMENTAL PARTICLES CLASSIFICATIONS.

8- ETC.,

IN EARTH BIOFRIENDLY ELECTRIC FUNDAMENTAL PARTICLES CLASSES ARE IN CHARGE OF OPERATIONS OF ALL EARTHS LIVING THINGS BODY ORGANS AND BODY SYSTEMS OPERATIONS AS WELL AS ARE IN CHARGES OF CENTRAL AND PERIPHERAL INTELLIGENCE SYSTEMS CENTERS OPERATIONS, WITHOUT ELECTRIC BIOFRIENDLY FUNDAENTAL PARTICLES LIFE CAN NOT EXIST.

BIOFRIENDLY LIGHT FUNDAMENTAL PARTICLES, BIOFRIENDLY SONIC FUNDAMENTAL PARTICLES AND LIGHT-SONIC FUNDAMENTAL PARTICLE INFORMATION IMAGE PARTICLE CLOUDS (S.Y. – F.P. – I.I. – P.cl.) ARE IN CHARGE OF PRODUCTIONS OF THOUGHT CURRENTS AND PSYCHE IN LIVING THINGS CNS, AS WELL AS MOST OF CNS INTERNAL ELECTRONS NUCLEON PARTICLE COMPOUNDS CONSTRUCTIONS ARE CONSTRUCTED FROM S.Y.- F.P. – I.I. – P.cl. COMPOUNDS, WITHOUT SONIC PARTICLE CLOUDS AND WITHOUT LIGHT PARTICLE CLOUDS AND THERMAL-ELECTRIC PARTICLE CLOUDS PSYCHE AND THOUGHT CURRENTS COULD NOT DEVELOP AND CAN NOT EXIST AT ALL.

OTHER CLASSES FUNDAMENTAL PARTICLE IN EARTH

THERE ARE OTHER PARTICLE CLASSES, SOME ARE AIRBORN BIOFRIENDLY PARTICLES AND OTHERS ARE BIOHOSTILE PARTICLES CONSTRUCT DIFFERENT LIVE OR NON-LIVE ELECTRON-NUCLEON PARTICLE COMPOUND CONSTRUCTIONS IN EARTH.

1 – AIRBORN INFRA- RED FUNDAMENTAL - PARTICLES (Y. i. r. – F.P. = T. i. r. – F.P.) WHICH CAN TRAVEL AND TRANSMIT THROUGH OTHER MEDIA ROUTES AS WELL, THESE PATICLES CLASSIFIED UNDER LIGHT PARTICLES ALSO IN FACT ARE THERMAL FUNDAMENTAL PARTICLE AS WELL.

INFRA-RED FUNDAMENTAL PARTICLES ARE THERMAL ENERGY CONTENTS PARTICLES, INFRA RED CAN PROVIDE AND DELIVER QUANTUM THERMAL ENERGY FOR ANY GIVEN NANO-LOCATIONS INSIDE ELECTRONS – NUCLEONS WHO ARE IN NEEDS OF THERMAL – ENERGIES TO PERFORM NANO-TASKS, OR TRIGGERING START BIOLOGICAL CHEMICAL INTERACTIONS INSIDE ELECTRONS-NUCLEONS SUBSYSTEM –UNITS CHEMICAL LAB.S WHO ARE IN NEED OF QUANTUM DOSE TEMPERATURES TO INITIATE A. S. I. P. F. Mol. C.I.C., OR THROUGH PROVIDING QUANTUM HEAT ENERGIES FOR ANY GIVEN QUANTUM –LOCATIONS INFRA RED CAN TRIGGER START AND ACHIEVE DIFFERENT THERMO-DYNAMI1C PHYSICAL TASKS, OR DO OTHER THERMAL – ELECTRICAL MIXED TASKS (QUANTUM THERMO-ELECTRIC TASKS) INSIDE ATOMS, CELLS, ETC.

2 - FROM OTHER AIRBORN FUNDAMENTAL PARTICLE CLASSES OF PLANET EARTH ARE ULTRA-VIOLET FUNDAMENTAL PARTICLE (Y. u. v. – F.P. = E. u. v. – F.P.) CLASS, ULTRAVIOLET IS WEAK ELECTRIC FUNDAMENTAL PARTICLES, E. u. v. – F.P. BELONG TO ELECTRICAL FUNDAMENTAL PARTICLES CLASSES OF EARTH.

ULTRAVIOLET IS THE WEAKEST ELECTRIC FUNDAMENTAL

PARTICLE IN PLANET EARTH, IT IS LOCATED THERE AFTER THE LIGHT FUNDAMENTAL PARTICLE CLASSES OF PLANET EARTH, ONE ULTRAVIOLETS FUNDAMENTAL PARTICLES ELECTRIC ENERGY CONTENT IS EQUAL TO ONE-UNIT ELECTRIC ENERGY, (THE QUANTITY OF ELECTRIC –ENERGY IN ONE ULTRAVIOLET PARTICLE IS EQUAL TO ONE UNIT ELECTRIC-ENERGY).

ULTRA VIOLET PARTICLES ARE ELECTRIC ENERGY PROVIDERS FOR QUANTUM SITES INSIDE ELECTRONS NUCLEONS AND CAN BE USED AS ELECTRIC ENERGY PROVIDER FOR INTERNAL ELECTRON NUCLEON NANO-LOCATIONS FOR ACHIEVING INTERNAL ELECTRON- NUCLEON NANO – WORKS OF DIFFERENT KINDS.

3 - MOST OF EARTHS PARTICLES EITHER BIOFRIENDLY OR BIOHOSTILE PRESENTLY HAVE NOT BEEN DISCOVERED AND NOT KNOWN YET.

THE NON DISCOVERED BIOFRIENDLY FUNDAMENTAL PARTICLE ARE BUILDING BLOCK OF ELECTRONS – NUCLEONS PARTICLE COMPOUND CONSTRUCTIONS AND ARE CREATORS WHO CONSTRUCTED MOST OF DIFFERENT LIVE ELECTRONS-NUCLEONS PARTICLE COMPOUND CONSTRUCTIONS.

EARTHS BIOFRIENDLY SENSIBLE SONIC FUNDAMENTAL PARTICLES AND OTHER CLASS PARTICLES
---
1 - PLANET EARTHS SONIC FUNDAMENTAL PARTICLES MOSTLY ARE LESS - WEIGHT LESS ENERGY AND LESS GRAVITON

BIOFRIENDLY BENIGN WEAK SOUND FUNDAMENTAL PARTICLES (S. – F.P.), THE SONIC FUNDAMENTAL PARTICLES INFORMATION – IMAGE – PARTICLE –CLOUDS (S. –F.P. – I.I.- P.cl.) CONSTRUCT MOST OF INTERNAL ELECTRON-NUCULEONS CNS- PARTICLE COMPOUND MOLECULAR CONSTRUCTIONS AT AUDITARY CNS CENTERS ATOMS OF BRAIN CELLS.

2 - ADDITIONALLY, THE SONIC – LIGHT FUNDAMENTAL PARTICLE INFORMATION IMAGE PARTICLE CLOUDS (S. Y. – F.P. – I.I. – P. cl.) JOINTLY CONSTRUCT THOUGHT CURRENTS AND PSYCHE AT MOST PLANTS AND ANIMAL SPECIES, THERE ARE FEW OTHER BIOFRIENDLY EARTHS FUNDAMENTAL PARTICLES SUCH AS THERMAL FUNDAMENTAL PARTICLES (T. – F.P.) BIOFRIENDLY ELECTRIC FUNDAMENTAL PARTICLES (E. – F.P.) ALSO PARTICIPATE IN PSYCHE – GENESIS PHENOMENON, NEO- GENESIS OF INTERNAL ELECTRON- NUCLEON PARTICLE COMPOUNDS CONSTRUCTIONS THROUGH USE OF MOLECULAR S. Y. – F.P. – I.I. – P.cl. _ COMPOUND CONSTRUCTIONS ALSO ARE F.P. –I.I.- P.cl. STRORAGE PHENOMENON AS WELL.

3 – IN EARTHS ATMOSPHERE THERE IS NOT AIRBORN FREE HIGH ENERGY BIOHOSTILE SONIC FUNDAMENTAL PARTICLES, THIS WAS ALSO A FACTOR HELPED FOR SYNTHESIS OF LIVING THINGS IN EARTH, WEAK BIOFRIENDLY SONIC PARTICLE COMPOUNDS ARE USED IN CONSTRUCTION OF EARTHS ELECTRONS NUCLEONS.

4 - SUPERSONIC FUNDAMENTAL PARTICLES, MICROWAVE PARTICLES MOSTLY HAVE BEEN USED IN COMMERCIAL,

MEDICAL OR NON MEDICAL INDUSTRIAL AND TELECOMMUNICATION FIELDS, THESE FUNDAMENTAL PARTICLES AS WELL AS THERE ARE LARGE OTHER UNKNOWN SONIC FUNDAMENTAL PARTICLE CLASSES IN EARTH WHICH ARE NON SENSIBLE TO MAN KIND SPECIES, STILL SOME OF THOSE SONIC PARTICLES ARE SENSIBLE TO ANOTHER ANIMALS.

5 - NON DISCOVERED FUNDAMENTAL PARTICLES NUMBERS IN EARTH FAR EXCEED FEW ABOVE MENTIONED KNOWN FUNDAMENTAL PARTICLE CLASSES, THERE ARE FAR LARGE NUMBERS OF OTHER BIOFRIENDLY FUNDAMENTAL PARTICLES ARE NON SENSIBLE AND NOT DISCOVERED BIOHOSTILE PARTICLES IN EARTH, AND ARE USED IN PARTICLE – COMPOUNDS CONSTRUCTING OF PLANET EARTHS DIFFERENT ATOMS ELECTRONS NUCLEONS PARTICLE COMPOUND CONSTRUCTIONS.

6 – DESCRIPTION OF LARGE NUMBERS OF DIFFERENT BIOHOSTILE FUNDAMENTAL PARTICLES CLASSES SUCH AS, X-PARTICLES CLASS, GAMMA PARTICLES CLASSES, LASER, INDUSTRIAL ELECTRIC PARTICLES, HARSH INDUSTRIAL THERMAL PARTICLES CLASSES HAS NO USE FOR THIS BOOK.

NANO-UNIT'S ENERGY SOURCES,
AND ENERGY PROVIDER TO NANO - LOCATIONS
QUANTUM ENERGY PROVIDERS TO PERIPHERON'S,
AND NUCLEUS'S NANO LOCATIONS

PERFORMING PHYSICAL, CHEMICAL BIOLOGICAL
TASKS, INSIDE ELECTRONS, NUCLEONS.

THERMAL FUNDAMENTAL PARTICLES STORE HEAT –ENERGY IN FORMS OF THERMAL PARTICLE COMPOUNDS INSIDE ELECTRONS- NUCLEONS AND PROVIDE THERMAL ENERGIES FOR NANO –LOCATIONS WHEN IT IS NEEDED, ELECTRIC PARTICLES PROVIDE ELECTRICITY ENERGY IN QUANTUM AMOUNTS TO BE USED INSIDE DIFFERENT QUANTUM LOCATIONS FOR ACHIEVING DIFFERENT KIND NANO-TASKS INSIDE DIFFERENT NANO-UNITS, LIGHT FUNDAMENTAL PARTICLES SONIC PARTICLES AND ALL OTHER FUNDAMENTAL PARTICLES DO THE SAME.

UNITS OF THERMAL – ENERGY
INFRA- RED THERMAL FUNDAMENTAL PARTICLE

DEFINITION: THE THERMAL ENERGY CONTENT OF ONE INFRA- RED PARTICLE IS EQUALS TO ONE THERMAL – ENERGY-UNIT,

THE EXISTING THERMAL ENERGY AMOUNT IN ONE INFRA- RED THERMAL FUNDAMENTAL PARTICLE IS EQUAL TO ONE UNIT THERMAL ENERGY.

INFRA – RED FUNDAMENTAL PARTICLE (T. i. r. – F.P.) IS A BIOFRIENDLY WEAK THERMAL FUNDAMENTAL PARTICLE IN EARTH, THE INFRA-RED SIMILAR TO OTHER THERMAL PARTICLES PROVIDES QUANTUM THERMAL ENERGY INTO INTERNAL ELECTRON NUCLEON QUANTUM LOCATIONS (NANO-SITES) TO BE USED TO PERFORM DIFFERENT BIO- LOGICAL CHEMICAL AND PHYSICAL NANO-TASKS, THROUGH THE USE OF PARTICLE PROVIDED THERMAL ENERGIES.

INFRA RED FUNDAMENTAL PARTICLE COMPOUNDS (T. i. r. – F. P. _ COMP.) SIMILAR TO OTHER PARTICLES HAVE BEEN USED IN CONSTRUCTIONS OF DIFFERENT INTERNAL ELEC- TRONS-NUCLEONS PARTICLE COMPOUNDS STRUCTURES (THIS IS PHENOMENON OF THERMAL –ENERGY STORAGE PHENOMENON INSIDE PARTICLE COMPOUNDS) AND UN- DER DEGENERATIVE A. S. I. F.P. Mol. C.I.C. THESE PARTICLE COMPOUNDS CAN RELEASE INFRA RED PARTICLES FREE IN INTERNAL ELECTRON-NUCLEON NANO-SITES, AS THERMAL ENERGY PROVIDER IN ANY GIVEN HEAT ENERGY NEEDED NANO-LOCATIONS.

BIOFRIENDLY RED COLOR FUNDAMENTAL PARTICLE

RED COLOR LIGHT FUNDAMENTAL PARTICLE (Y. r. –F.P.) OR RED COLOR THERMAL FUNDAMENTAL PARTICLE (T. r. – F.P.), THE RED COLOR FUNDAMENTAL PARTICLES POSSESSES TWO DIFFERENT KIND BIOFRIENDLY THERMAL ENERGY, AND BIOFRIENDLY LIGHT ENERGIES.

THE INTERNAL ELECTRON'S - NUCLEON'S RED LIGHT PARTICLE COMPOUNDS ARE QUANTUM THERMAL ENERGY STORAGE SYSTEMS, AND QUANTUM ENERGY PROVIDERS BECAUSE OF THEIR INHERITANT ENERGY CONTENTS.

BIOFRIENDLY ELECTRIC ENERGY PROVIDER PARTICLES TO NANO-SITES

INTERNAL ELECTRON NUCLEON ENERGY STORAGE SYSTEMS AND ENERGY PROVIDER SYSTEMS

BLUE COLOR PARTICLES AND VIOLET COLOR PARTICLES ARE ELECTRIC ENERGY DONNOR PARTICLES

DIFFERENT FUNDAMENTAL PARTICLES POSSESSES DIFFERENT KIND ENERGY QUALITIES AND QUANTITIES, FOR EXAMPLE THE THERMAL ENERGY CONTENTS OF DIFFERENT FUNDAMENTAL PARTICLES AT DIFFERENT CLASSES THERMAL FUNDAMENTAL PARTICLES VARY REMARKABLY FROM EACH OTHER, WHEN COMPARING ONE PARTICLE CLASS

FUNDAMENTAL PARTICLES ENERGY CONTENT WITH ANOTHER CLASS PARTICLES ENERGIES, THE RESULTS ARE DIFFERENT BUT ALWAYS THE FINDINGS ARE FUNDAMENTAL PARTICLE SPECIFIC.

SOME FUNDAMENTAL PARTICLES ARE MULTI – ENERGY PROVIDERS AND POSSESSES DIFFERENT KIND ENERGIES CONTENTS, THE BIOFRIENDLY RED COLOR FUNDAMENTAL PARTICLES POSSESSES TWO KIND THERMAL ENERGY AND LIGHT ENERGY BOTH, THE RED COLOR FUNDAMENTAL PARTICLES ARE THERMAL ENERGY PROVIDERS AS WELL AS LIGHT –ENERGY DONNOR'S, THE RED COLOR PARTICLES COMPOUNDS ALSO ARE THERMAL-ENERGY AND LIGHT ENERGY STORAGE SYSTEMS INSIDE THE ELECTRONS NUCLEONS PARICLE COMPOUND CONSTRUCTIONS.

THE VIOLET COLOR FUNDAMENTAL PARTICLES AND THE BLUE COLOR FUNDAMENTAL PARTICLES FROM PLANET SUN ORIGIN, BOTH ALSO ARE MULTI – ENERGY TWO DIFFERENT KIND FUNDAMENTAL PARTICLES AS WELL, BLUE COLOR FUNDAMENTAL PARTICLES AND VIOLET COLOR FUNDAMENTAL PARTICLES BOTH POSSESSES ELECTRIC- ENERGIES AS WELL AS LIGHT ENERGIES IN PARTICLE CONTENT, BLUE PARTICLES AND VIOLET PARTICLES ARE MULTI - ENERGY CONTENT PARTICLES FROM TWO DIFFERENT KIND PARTICLES CLASSES.

BOTH OF BLUE COLOR PARTICLES AND VIOLET COLOR PARTICLES ARE ELECTRIC- ENERGY PROVIDERS PARTICLES AS WELL AS LIGHT ENERGY DONNOR'S PARTICLES, BOTH BLUE PARTICLES AND VIOLET PARTICLES CAN GET INTO DIFFERENT

NANO-LOCATIONS OF DIFFERENT ELECTRONS NUCLEONS AND PROVIDE NEEDED ELECTRIC ENERGIES OR LIGHT ENERGIES, TO ANY GIVEN NANO—LOCATIONS WHO ARE IN NEED OF ELECTRIC OR LIGHT ENERGIES TO DO QUANTUM NANO TASKS THROUGH USE OF THOSE PROVIDED ELECTRIC-LIGHT ENERGIES.

ENERGY SOURCES AND PROVIDERS FOR NANO TASKS

INSIDE ELECTRONS NUCLEONS, THE
PARTICLES AT QUANTUM LOCATIONS

FUNDAMENTAL PARTICLES CONTROL CHEMICAL, PHYSICAL, BIOLOGICAL, TASKS THROUGH PROVIDING, OR STOPPING ENERGY SOURCES, TO DIFFERENT NANO- SITES FOR NANO-FUNCTIONS, INCLUDING A. S. I. F.P. Mol. C.I.C.

PHENOMENON OF ENERGY STORAGE- PARTICLES

PHENOMENON OF ENERGY STORAGE INSIDE
ELECTRONS NUCLEONS, THE PARTICLES

FUNDAMENTAL PARTICLES NATURALLY POSSESS INHERITANT FUNDAMENTAL PARTICLE SPECIFIC ENERGIES, WEIGHT AND GRAVITON. FUNDAMENTAL PARTICLES TRANSIT THROUGH DIFFERENT PARTICLE TRANSMISSION ROUTES INTO INSIDE ELECTRONS NUCLEONS, AND UNDER REGENERATIVE A. S. I. F.P. Mol. C.I.C. COMBINE WITH INTER ELECTRON NUCLEON PARTICLE COMPOUNDS AND CONSTRUCT

ELECTRONS NUCLEONS PARTICLE- COMPOUND CONSTRUCTIONS, THEREFORE FUNDAMENTAL PARTICLES STORE THEIR ENERGY CONTENT IN PARTICLE COMPOUND FORMS INSIDE ELECTRONS NUCLEONS, THIS IS PHENOMENON OF STORAGE OF ENERGY INSIDE ELECTRONS NUCLEONS.

PHENOMENON OF ENERGY RELEASE FROM NANO-UNITS TO DIFFERENT NANO-SITES BY PARTICLES PHENOMENON OF ENERGY RETRIEVAL FROM ELECTRONS-NUCLEONS TO OUTSIDE

FUNDAMENTAL PARTICLES PRESENCE IN ANY GIVEN QUANTUM LOCATIONS INSIDE ELECTRONS NUCLEONS IS EQUAL TO EXISTENCES OF FUNDAMENTAL PARTICLE ENERGY IN THAT GIVEN NANO-SITE AS WELL, FUNDAMENTAL PARTCLES IN ANY GIVEN NANO-LOCATIONS CAN PROVIDE NEEDED ENERGY AND INITIATE QUANTUN PHYSICAL CHEMICAL OR BIOLOGICAL FUNCTIONS THROUGH THE USE OF EXISTING PARTICLE ENERGY AND PROVIDING PARTICLE ENERIES TO ACCOMPLISH NEEDED CHEMICAL BIOLOGICAL OR PHYSICAL NANO-TASK ACCORDING THE PARTICLE MOLECULAR EVOLUTION ORDERS AND PATHS.

UNDER DEGENERATIVE A. S. I. F.P. Mol. C.I.C. THE CONSTRUCTED FUNDAMENTAL PARTICLES – COMPOUNDS BREAK DOWN INTO SMALLER FUNDAMENTAL PARTICLES FORMS AND RELEASE PARTICLE ENERGY INTO NANO-LOCATION, FUNDAMENTAL PARTICLES POSSESS ENERGY AND PROVIDE ENERGY INTO RELEASED QUANTUM LOCATIONS AS ENERGY DONNOR'S PROVIDE THEIR ENERGIES AND

THROUGH USE OF THAT ENERGY ACHIEVE CHEMICAL PHYSICAL OR ANY OTHER NEEDED TASKS IN QUANTUM LOCATIONS THROUGH USE OF RELEASED PARTICLES ENERGIES.

IN OTHER SITUATIONS, WHEN RELEASED AND FREED FUNDAMENTAL PARTICLES LEAVE NANO SITES AND SCAPE TO OTHER LOCATIONS FROM NANO-SITES AS FREE PARTICLES, IN CASE OF ABSENCE OF FUNDAMENTAL PARTICLES TO ANOTHER LOCATIONS BY ITSELF CREATE ENERGY VACUUM AND NO ENERGY IN NANO-LOCATIONS, THEREFORE THE PHYSICAL CHEMICAL BIOLOGICAL FUNCTION ALL STOP THROUGH SOURCE CUT, IN THE ABCENSE OF FUNDAMENTAL PARTICLES NANO-TASKS STOP ACCORDINGLY.

THIS PHENOMENON SIMILAR TO ON-OFF KEY, CONTROL AUTONOMOUS SEQUENTIAL INTERACTIONS, START OR STOP NANO TASKS IN NANO-LOCATIONS, THROUGH PROVIDING FUNDAMENTAL PARTICLE ENERGY INTO GIVEN QUANTUM LOCATION TRIGGER START NANO-FUNCTIONS AUTONOMOUSLY, IN THE ABSENCE OF DONOR FUNDAMENTAL PARTICLE ENERGY THE AUTONOMOUS SEQUENCES TRIGGER STOP- FUNCTIONS IN NANO-LOCATIONS, IN START PROVIDE ENERGY AND IN STOP INTERRUPT ENERGY SOURCES.

RED COLOR PARTICLES OPERATED THERMAL – FUNCTIONS

UNDER REGENERATIVE A. S. I. F.P. Mol. C.I.C. THE RED COLOR FUNDAMENTAL PARTICLES INTER INTO CELLS ELECTRONS-NUCLEONS SUBSYSTEM UNITS COMBINE WITH PARTICLE COMPOUNDS AND PRODUCE RED COLOR PARTICLE

COMPOUND CONSTRUCTIONS, AS RED COLOR PARTICLES ENERGY MOSTLY IS THERMAL ENERGY AS WELL AS LIGHT ENERGY, IN RESULT RED- PARTICLE COMPOUND CONSTRUCTION IS THERMAL –ENERGY STORAGE SYSTEM AS WELL, THIS PHENOMENON IS THERMAL –ENERGY STORAGE AND LIGHT-ENERGY STORAGE SYSTEMS PHENOMENON.

RED- PARTICLE- COMPOUND CONSTRUCTIONS OF ELECTRONS NUCLEONS ARE THERMAL – LIGHT ENERGIES STORAGE SYSTEMS, WHICH THROUGH DEGENERATIVE AUTONOMOUS SEQUENTIAL CHEMICAL INTERACTIONS CYCLES (DEG. A. S. I. F.P. Mol. C.I.C.) RED THERMAL PARTICLES CAN BE RELEASED IN ANY INTERNAL ELECTRON- NUCLEON NANO SITE PROVIDE ENERGY IN EXACT NEEDED QUANTUM-LOCATIONS OR THE RELEASED PARTICLES CAN INTERACT WITH OTHER PARTICLES AND PERFORM INNORMOUS DIFFERENT KINDS PHYSICAL CHEMICAL BIOLOGICAL FUNCTIONS, ALSO RELEASED PARTICLES PARTICIPATE IN INNORMOUS DIFFERENT KIND A. S.I. F.P. Mol. C.I.C..

THE STORED RED-COLOR PARTICLES WHEN RELEASED IN NANO-SITE PROVIDE THERMAL –ENERGY FOR LIVING THINGS INTERNAL ELECTRON – NUCLEON INTERNAL SUBSYSTEM-UNITS CHEMICAL LAB.S TO TRIGGER START AND CONDUCT A. S. I. F.P. Mol. C.I.C., ADDITIONALLY WHEN RED COLOR PARTICLE ALL CONSUMED AND NOT PRESENT IN QUANTUM LOCATIONS ANY MORE AND NO MORE THERMAL ENERGIES PRESENT, UNDER THESE CONDITIONS THE A.S. I. F.P. Mol. C.I.C. STOPS IMMEDIATELY AS WELL, THE RED COLOR THERMAL PARTICLES OPERATED BIOLOGICAL

PHYSICAL CHEMICAL NANO-TASKS THROUGH PROVIDING THERMAL ENRGIES AND LIGHT ENEGIES AS WELL.

INFRA- RED AND RED COLOR FUNDAMENTAL PARTICLES AS WELL AS NUMEROUS OTHER NON- DISCOVERED THERMAL PARTICLES ALL TOGETHER WITH COOPERATIONS OF EACH OTHER PROVIDE QUANTUM THERMAL ENERGY FOR PERFORMING BIOLOGICAL CHEMICAL PHYSICAL NANO-TASKS AT INTERNAL ELECTRON- NUCLEON NANO – LOCATIONS OF LIVING THINGS CELLS.

AUTONOMOUS CONTROLS OF INTERNAL ELECTRON NUCLEON NANO TASKS BY PARTICLES

ALL OTHER FUNDAMENTAL PARTICLES SUCH AS ELECTRIC PARTICLES, SONIC PARTICLES, X-PARTICLES, GAMMA PARTICLES, LIGHT PARTICLES, ETC., ALL FOLLOW THE SAME EXACT FUNDAMENTAL PARTICLE RULES IN OPERATIONS OF PHYSICAL CHEMICAL BIOLOGICAL TASKS IN INTERNAL ELECTRON NUCLEON NANO LOCATIONS SIMILAR TO RED COLOR FUNDAMENTAL PARTICLES AS EXPLAINED IN ABOVE.

PROVIDING NEEDED ANY KIND ENERGIES SUCH AS ELECTRIC, LIGHT, SONIC, X, GAMMA ENERGIES, ETC., IMMEDIATELY TRIGGER START DIFFERENT PHYSICAL CHEMICAL BIOLOGICAL NANO-TASKS IN GIVEN INTER ELECTRON NUCLEON NANO-LOCATIONS AND ACCOMPLISH INTENDED QUANTUM TASK AND AT THE END, THE FUNDAMENTAL PARTICLES WHICH WAS PROVIDING DIFFERENT ELECTRIC-LIGHT- SONIC- GAMMA- X- ENERGIES TERMINATE AND TRIGGER STOP

DIFFERENT ONGOING PHYSICAL CHEMICAL BIOLOGICAL NANO FUNCTIONS THROUGH LEAVING SCAPING AREA OR WHEN CONSUMED ALL IN DIFFERENT CHEMICAL PHYSICAL BIOLOGICAL INTERACTIONS THE END OF ENERGY SOURCE IS EQUAL END OF FUNCTIONS, IMMEDIATELY TERMINATE THE CONTINUATIONS OF THE ONGOING TASKS.

ALL FUNDAMENTAL PARTICLES PHYSICAL CHEMICAL BIOLOGICAL INTERACTIONS TAKE PLACE WITH EXPONENTIAL SPEED AUTONOMOUSLY AND SEQUENTIALLY, ALL INTERACTIONS EITHER IN LIVE OR NON-LIVE NANO-UNITS ALL FOLLOW THE SAME FUNDAMENTAL PARTICLE RULES AND LAWS ACCROSS THE UNIVERSE.

GREEN COLOR LIGHT FUNDAMENTAL PARTICLES (Y. g. – F.P.) QUANTUM LIGHT ENERGY POVIDER FUNDAMENTAL PARTICLES UNIT OF LIGHT ENERGY

DEFINITION: THE EXISTING LIGHT ENERGY AMOUNT IN ONE GREEN LIGHT FUNDAMENTAL PARTICLE IS EQUAL TO ONE UNIT LIGHT- ENERGY.

GREEN LIGHT FUNDAMENTAL PARTICLE (Y. g. – F.P.) IS AN ESSENTIAL BIOFRIENDLY DOMINANT LIGHT FUNDAMENTAL PARTICLE ON EARTH.

LIGHT ENERGY –CONTENT OF ONE GREEN COLOR LIGHT PARTICLE (Y. g. –F.P.) IS EQUAL TO ONE UNIT LIGHT – ENERGY- AMOUNT,

THE DIFFERENT LIGHT FUNDAMENTAL PARTICLES HAVE DIFFERENT LIGHT – ENERGY CONTENTS, DIFFERENT PARTICLE WEIGHTS, DIFFERENT GRAVITON AS WELL AS DIFFERENT PHYSICAL CHEMICAL BIOLOGICAL PROPERTIES, THE LIGHT ENERGIES FROM DIFFERENT SOURCES CAN BE QUANTIZED AND MEASURED IN COMPARE TO ONE UNIT LIGHT-ENERGY THAT DEFINED IN ABOVE, THE YELLOW COLOR LIGHT

FUNDAMENTAL PARTICLE (Y. y.- F.P.) IN EARTH IS ANOTHER EXAMPLE FROM SENSIBLE BIOFRIENDLY LIGHT PARTLE CLASSES, THERE ARE LARGE NUMBERS OF OTHER NON DISCOVERED LIGHT FUNDAMENTAL PARTICLE CLASSES IN EARTH WHICH HAVE BEEN USED IN CONSTRUCTIONS OF DIFFERENT NANO-UNITS.

UNIT OF ELECTRIC – ENERGY
ELECTRIC ULTRAVIOLET FUNDAMENTAL PARTICLE (E. u. v. – F.P.)

DEFINITION: THE EXISTING ELECTRIC ENERGY AMOUNT IN ONE ULTRAVIOLET ELECTRIC PARTICLE IS EQUALTO ONE UNIT OF ELECTRIC ENERGY.

AN ULTRAVIOLET FUNDAMENTAL PARTICLE (E. u. v. – F.P.) IS THE WEAKEST ELECTRIC FUNDAMENTAL PARTICLE BETWEEN EARTH'S ELECTRIC FUNDAMENTAL PARTICLES, AN ULTRAVIOLET PARTICLE POSSESSES THE LEAST PARTICLE WEIGHT, PARTICLE GRAVITON AND ELECTRIC ENERGY CONTENT IN COMPARISON TO OTHER EARTHS ELECTRIC PARTICLES, THE ELECTRIC ENERGY AMOUNT OF ONE ULTRAVIOLET FUNDAMENTAL PARTICLE IS EQUAL TO ONE UNIT ELECTRIC ENERGY QUANTITY, THE ULTRAVIOLET PARTICLE IS LOCATED AFTER FINISH LINES OF LIGHT FUNDAMENTAL PARTICLE CLASS AND BEFORE THE STRATING POINTS OF HARSH BIOHOSTILE HIGH ENERGY INDUSTRIAL ELECTRIC FUNDAMENTAL PARTICLE CLASSES.

QUANTUM ELECTRIC – ENERGY PROVIDERS FOR INTER ELECTRON-NUCLEON NANO- SITES OF LIVING THINGS

VIOLET COLOR ELECTRIC PARTICLES (E. v. – F.P.), BLUE COLOR ELECTRIC PARTICLES (E. b. – F.P.)

BLUE COLOR LIGHT PARTICLES (Y. b. – F.P.) AND VIOLET COLOR LIGHT PARTICLE (Y. v. – F.P.)

VIOLET COLOR PARTICLES (Y. v. – F.P.) AND BLUE COLOR FUNDAMENTAL PARTICLES (Y. b. – F.P.) POSSESSES MIXED LIGHT AND ELECTRIC ENERGIES, BOTH PARTICLES ARE LIGHT ENERGY AND ELECTRIC ENERGY PROVIDER INTO INTERNAL ELECTRON - NUCLEON NANO-SITES TO BE USED FOR ACHIEVING QUATUM BIOLOGICAL PHYSICAL CHEMICAL FUNCTIONS, IN EARTH THERE ARE LARGE OTHER UNKNOWN NON SENSIBLE MULTI ENERGY PROVIDER FUNDAMENTAL PARTICLES, WHICH CONSTRUCT DIFFERENT INTERNAL ELECTRON – NUCLEON PARTICLE COMPOUNDS CONSTRUCTIONS OF ATOMS AND PRESENTLY REMAIN UNKNOWN.

BLUE COLOR PARTICLES, VIOLET COLOR PARTICLES ARE BIO-FRIENDLY LIGHT - ELECTRIC PARTICLES, IN LIVING THINGS THESE PARTICLES PROVIDE ELECTRIC ENERGIES AND LIGHT ENERGIES INTO INTERNAL ELECTRON NUCLEON QUATUM LOCATIONS, WITH COOPERATION OF OTHER BIOFRIENDLY FUNDAMENTAL PARTICLES OPERATE ENTIRE LIVING THINGS BIOLOGICAL, CHEMICAL, PHYSICAL NANO FUNCTIONS IN DIFFERENT QUANTUM LOCATIONS.

HOW FUNDAMENTAL PARTICLES OPERATE LIVING THINGS
NANO NEUROLOGY OF BODY ORGANS
CENTRAL INTELLIGENCE SYSTEM
CENTERS OF BODY ORGANS (CIS)

FUNCTIONS OF BODY ORGANS ELECTRONS NUCLEONS POPULATIONS CONTROLLED BY ORDERS OF ORGANS CENTRAL INTELLIGENCE SYSTEMS.

ALL BODY SYSTEMS SUCH AS G.I.S., G.U.S., CARDIO-PULMONARY SYSTEMS, ETC. ALL HAVE SYSTEM SPECIFIC CENTRAL INTELLIGENCE SYSTEM CENTERS AS WELL AS SYSTEM SPECIFIC PARTICLE CIRCULATION SYSTEM, ABOVE FACTS ALSO ARE TRUE IN REGARD TO BODY ORGANS WHICH EACH GIVEN BODY ORGAN POSSESSES ORGAN SPECIFIC CENTRAL INTELLIGENCE SYSTEM AS WELL AS ORGAN SPECIFIC PARTICLE CIRCULATION SYSTEMS,THESE PARTICLE CIRCULATION SYSTEMS AND ORGAN CENTRAL INTELLIGENCE SYSTEMS MOSTLY ARE NANO PARTICLE CIRCULATION SYSTEMS AND NANO CENTRAL INTELLIGENCE SYSTEM CENTERS, PRESENTLY KNOWN ANATOMICAL OR HISTOLOGICAL NERVOUS SYSTEMS ARE NOT NANO STRUCTURES OR QUANTUM CONSTRUCTIONS.

CIRCULATION OF NORMAL PARTICLE CLOUDS WILL PRODUCE NORMAL FUNCTIONAL ORGANS AND SYSTEMS, IN THE CASES OF ABNORMALITY IN PARTICLE CLOUD CIRCULATIONS OR ABNORMAL F.P. − I.I. − P. cl. CIRCULATIONS WILL CAUSE FUNCTION DISORDERS AND IS THE CAUSE FOR LARGE NUMBERS OF IDIOPATHIC DISORDERS SUCH IRRITABLE G.I.S. DISORDERS, RESPIRATORY SYSTEMS AND G. U. S. DISORDER, NEUROSIS, ETC.

IN ANIMALS, EACH GIVEN ORGAN CONSTRUCTED FROM THREE DISTICT STRUCTURES: 1 − CIS, 2- PCS, 3 − NANO − UNIT STRUCTURES OF DIFFERENT ORGANS, MALFUNCTION OF ANY STRUCTURE CAN CAUSE MALFUNCTION AND ORGAN DISORDER.

1 - CENTRAL INTELLIGENCE SYSTEM CENTERS OF ORGANS (CIS):

THE CENTRAL INTELLIGENCE SYSTEM OF ORGANS ARE IN CHARGE OF ORGANS ELECTRONS NUCLEONS, PHYSICAL, CHEMICAL, BIOLOGICAL FUNCTIONAL OPERATIONS, AND MOST OF PHYSICAL CHEMICAL BIOLOGICAL FUNCTIONS OF ORGANS ELECTRON NUCLEON OPERATIONS CONTROLLED THROUGH BIOFRIENDLY PARTICLE CLOUDS AND ELECTRIC FUNDAMENTAL PARTICLE CLOUD ORDERS OF ORGANS CENTERAL INTELLIGENCE SYSTEMS OR THE PARTICLE CLOUD CURRENTS COMING FROM HIGHER OTHER INTELLIGENCE SYSTEM CENTERS AND PRODUCING EITHER NORMAL OR ABNORMAL FUNCTIONS ORGANS, CIS CURRENTS FLOWING THROUGH PARTICLE CIRCULATIONS SYSTEMS TO ENTIRE

ELECTRON NUCLEON POPULATION OF ANY ORGAN, END RESULTS IS NORMAL OR ABNORMAL ORGAN FUNCTION.

FOR EXAMPLE, THE PARTICLE CLOUDS REPORTS OF NITROGLYCERIN (UNDER TONGUE) THROUGH ORGANS CIS - PCS TRANSIT TO CARDIAC ORGAN CAUSE VASODILATION AND PAIN RELIEF, AT THIS PHENOMENON ORAL NITROGLYCERIN'S INFORMATION PARTICLE CLOUD TRANSIT THROUGH LIGHT SPEED PARTICLE CLOUD CURRENTS AND PRODUCE SYMPTON CHANGE BY PARTICLE CLOUD INFORMATION PRODUCED VASODILATION.

BUT IN MANY CASES MALFUNCTION OF PARTICLE CIRCULATION SYSTEMS OR ABNORMALITY OF CIRCULATING PARTICLE INFORMATIONS AND IMAGE CLOUDS PRODUCE LARGE NUMBERS OF FUNCTIONAL DISORDERS IN DIFFERENT ORGANS OF MOST BODY SYSTEMS AND ORGANS WHICH ARE KNOWN PRESENTLY AS IDIOPATHIC, PARTICLE CLOUD ABNORMALITIES OR PARTICLE CIRCULATION SYSTEM ABNORMALITIES MAY PRODUCE SIMPLE ORGAN IRRITABILITY SUCH AS IRRITABLE BOWL SYNDROMME OR FUNCTION ABNORMALITY SUCH AS DYSPNEA, OR NEUROSIS, BUT LATER ADDED FURTHER COMPLICATIONS SUCH AS INFECTIONS OR INFLAMMATORY ORGANIC SECONDARY DISEASES FURTHER COMPLICATE AND CHANGE FOR ANOTHER DISEASE PRODUCTIONS LIKE CROHN'S, OR DESTRUCTIVE SECONDARY LUNG DISEASES.

THE ABNORMALITY IN PARTICLE CLOUDS, F.P.- I.I. – P.cl. AS WELL ABNORMALITY OF PARTICLE CIRCULATION SYSTEMS

PRODUCE DISORDERS SUCH AS IRRITABLE BOWL, STOMACH, ETC. SYNDROMES, OR FUNCTIONAL DISORDERS IN BREATHINGS, SWALLOWING AND SIMILAR MANY OTHERS IN G. U. S. IRRITABLE DISORDERS WHICH AFTER PASSING TIMES MANY OTHER PERMANENT INFECTIONS OR INFLAMMATORY COMPLICATIONS ADD AND DEVELOPE FURTHER SERIOUS DISORDERS LIKE CROHN'S, ASTHMA, OR DESTRUCTIVE DISORDERS.

PRODUCTIONS OF ABNORMAL PARTICLE CLOUDS, ABNORMAL F.P. – I.I. – P.cl. BY CENTRAL INTELLIGENCE SYSTEM CENTERS, TRANSMISSION AND CIRCULATIONS OF THOSE ABNORMAL PARTICLE CLOUDS INSIDE ORGANS AND BODY SYSTEMS PARTICLE CIRCULATION SYSTEMS CIRCUITS PRODUCE ORGAN MALFUNCTIONS AND PRODUCE ABNORMAL SYMPTOMS AND SIGNS WHICH PATIENTS FEEL THOSE AND MOSTLY CALLED SOME THING IS WRONG IN BRAIN, IN MANY CASES THERE IS MALFUNCTION AT PARTICLE CIRCULATIONS DIFFERENT TYPES WHICH ARE IN OTHER VOLUMES,

2- FUNDAMENTAL PARTICLE CIRCULATION SYSTEM OF ORGANS (PCS):

FUNDAMENTAL PARTICLE CIRCULATION SYSTEMS OF BODY ORGANS MOSTLY ARE FLOWING ELECTRIC FUNDAMENTAL PARTICLES CURRENTS AND PARTICLE CLOUD CURRENTS OF ELECTRIC PARTICLES AS WELL OTHER CLASS FUNDAMENTAL PARTICLE CLOUD CURRENTS SUCH AS LIGHT PARTICLE CLOUDS, SONIC PARTICLE CLOUDS, THERMAL PARTICLES CURRENTS, ETC., WHICH THESE PARTICLE

CLOUDS PRODUCED AND ORIGINATED FROM CENTRAL INTELLIGENCE SYSTEM CENTERS, AND FROM THERE FLOW INSIDE NANO-NEURAL FIBERS PARTICLE CIRCULATION SYSTEMS DUCTS AS PARTICLE CURRENTS FINALLY INTERS INTO INTERNAL ELECTRONS - NUCLEONS POPULATIONS OF DIFFERENT ORGANS,

3- TOTAL BODY - ORGANS ELECTRON NUCLEON POPULATIONS.

CENTRAL INTELLIGENCE SYSTEM CENTERS OF BODY SYSTEMS (CIS), NANO NEUROLOGY OF BODY SYSTEMS:

NANO FUNCTIONS OF TOTAL ELECTRONS NUCLEONS POPULATIONS IN A BODY SYSTEM, IS UNDER NANO INTELLIGENCE CENTERS OF BODY SYSTEMS
FROM NANO-NEUROLOGICAL POINT OF VIEW EACH GIVEN BODY SYSTEM CONSTRUCTED FROM: 1 – C. I. S., 2 – P. C. S., 3 - NANO-UNIT STRUCTURES OF BODY SYSTEMS:

CENTRAL INTELLIGENCE CENTERS OF BODY SYSTEMS,

CENTRAL INTELLIGENCE CENTERS ORDER IN PARTICLE CLOUD FORMS, TRANSIT PARTICLE CLOUDS THROUGH PARTICLE CIRCULATIONS SYSTEMS TO DIFFERENT ORGANS ELECTRON NUCLEONS POPULATIONS, PARTICLE CLOUDS INTER INTO ELECTRONS, NUCLEONS AND UNDER A.S. I. F.P. Mol. C.I.C. INTERACT WITH EACH OTHER, ALL EXISTING ORGANS IN A GIVEN BODY SYSTEM FUNCTION UNDER BODY

SYSTEMS CENTRAL INTELLIGENCE CENTERS ORDERS PARTICLE CLOUD INSTRUCTIONS,

THE BODY SYSTEM'S CENTRAL INTELLIGENCE CENTERS PRODUCE AND REPORT F.P. – I.I. – P. cl. (PARTICLE CLOUDS) TO HIGHER CNS AND CENTRAL AUTONOMOUS SYSTEMS CENTERS AND FOLLOW THESE CENTERS ORDERS AND INSTRUCTIONS, THE C. I. S. ALSO ARE CONTROLLED THROUGH END NANO-UNIT FINDING REPORTS AND RECOMMENDATIONS, ALTHOUGH ARE ISSUERS OF ORDERS TO LOWER INTELLIGENCE CENTERS,

DIFFERENT BODY SYSTEMS MOSTLY THROUGH ADJACENT ADDITIONAL SIDE PARTICLE CIRCULATION SYSTEMS, CONNECT TO OTHER NEIGHBORING ORGANS AND SYSTEMS, THROUGH THESE SIDE PARTICLE CIRCULATION SYSTEMS CONNECTIONS THE SYMPTOMS AND SIGNS OF ONE SYSTEMS OR ONE ORGAN TRANSMIT AND SENSED AT NEIGHBORING SYSTEM OR ORGANS IN MANY EPISODES,

NANO –NEUROLOGY OF ANIMALS

PHENOMENON OF COMMUNICATIONS OF TOTAL BODY ELECTRONS NUCLEONS POPULATION WITH EACH OTHER BY PARTICLE CLOUDS.

F.P.- I.I. P. cl. EXCHANGE, AND PARTICLES CIRCULATION SYSTEMS BETWEEN TOTAL - BODY ELECTRONS NUCLEONS POPULATIONS.

HIERARCHY PHENOMENONS BETWEEN ELECTRONS NUCLEONS SOCIETY, THROUGH PARTICLE CLOUD EXCHANGE ORDERS.

HIERARCHY CELL SOCIETY SYSTEMS PHENOMENONS AND CELLS COMMUNICATION SYSTEMS

PHENOMENON OF COMMUNICATION BETWEEN TOTAL BODY ELECTRONS NUCLEONS, THROUGH PARTICLE CLOUD EXCHANGE SYSTEMS BETWEEN ELECTRONS NUCLEONS, AND FUNDAMENTAL PARTICLES CLOUD CIRCULATION SYSTEMS (P C S) THROUGH NANO NEURAL DUCT

CIRCULATION SYSTEMS IN ANIMALS DIVIDE INTO THREE MAJOR PARTICLE CIRCULATION SYSTEM (P. C. S.) CATEGORIES CLASSIFICATIONS:

1 – INDIGENOUS INTER - ELECTRON - NUCLEON F.P. –I.I.- P.cl. EXCHANGE, HIERARCHY SYSTEMS AND PARTICLE CIRCULATION SYSTEMS (I - PCS).

2- EXOGENOUS INTER –ELECTRON-NUCLEON PARTICLE CLOUD EXCHANGE, AND EXOGENOUS PARTCLE CIRCULATIONS SYSTEMS (EX - PCS).

3- CONNECTIONS AND PARTICLE CIRCULATION EXCHANGES BETWEEN: 1 & 2 (I - PCS & EX - PCS.).

TOTAL BODY PARTICLE CIRCULATING SYSTEMS (TB – PCS):

THIS IS PCS BETWEEN TOTAL BODY ELECTRONS, NUCLEONS POPULATION OF ANIMALS BODY, THIS IS THE FUNDAMENTAL PARTICLE CIRCULATION SYSTEMS THAT, CIRCULATE BETWEEN WHOLE-BODY ELECTRONS, NUCLEONS POPULATIONS INSIDE ANIMAL BODY, CONNECTS ALL BODY'S ATOMS TO EACH OTHER.

1 - INDIGENOUS PARTICLE CLOUD CIRCULATION SYSTEMS INSIDE ANIMAL BODY BETWEEN ELECTRONS NUCLEONS, AND INDIGENOUS INTER -ELECTRON - NUCLEON PARTICLE CLOUD EXCHANGES UNDER HIERARCHY SYSTEMS, THROUGH PARTICLE CLOUD EXCHANGE.

IN ANIMALS SPECIES, THE INDIGENOUS PARTICLE CIRCULATION SYSTEMS (I. PCS) ARE BI-DIRECTIONAL PARTICLE CLOUD TRANSPORTATION SYSTEMS, THE PCS CIRCULATE PARTICLE CLOUDS AND F.P. – I.I.- P. cl. BETWEEN TOTAL BODY ELECTRONS –NUCLEONS POPULATIONS UNDER HIERARCHY ORDER AND RESPOSE PARTICLE CLOUD CIRCULATION SYSTEMS.

THE PARTICLE CLOUD CIRCULATION SYSTEMS (PCS) TRANSMIT PERIPHERAL ELECTRONS NUCLEONS PARTICLE CLOUD INFORMATION AND IMAGES CURRENTS FROM PERIPHERAL ORGANS ELECTRONS NUCLEONS THROUGH NANO FIBER NEURAL DUCTS AND CARRY THOSE PARTICLE CLOUDS TO HIERARCHY HIGHER INTELLECTUAL CENTERS OF CNS AND AUTONOMOUS CENTER SYSTEMS ELECTRONS NUCLEONS REVIEWS AND DECISIONS.

THE HIGHER INTELLIGENCE SYSTEM CENTERS ELECTRONS NUCLEONS THROUGH DIFFERENT CENTER REVIEW A. S. I. F.P. Mol. C.I.C. PRODUCE ORDER - PARTICLE CLOUDS, THE ORDER PARTICLE CLOUDS THROUGH REVERSE DIRECTION PARTICLE CIRCULATION SYSTEM CURRENTS GET TRANSPORTIONS BACK TO PERIPHERAL ORGANS ELECTRONS NUCLEONS POPULATIONS AS INSTRUCTIONS FOR FUNCTION, WHICH UNDER THESE ORDER PARTICLE CLOUDS THE PERIPHERAL ORGAN ELECTRONS NUCLEONS FUNCTION ACCORDINGLY AND PERFORM NEEDED PHYSICAL CHEMICAL BIOLOGICAL NANO FUNCTIONS, ABOVE IS ONE CYCLE PARTICLE CIRCULATION SYSTEMS CYCLE WHICH TRANSPORT PARTICLE CLOUDS BETWEEN DIFFERENT LEVEL ELECTRONS NUCLEONS HIERARCHY PARTICLE CLOUDS ORDERS AND RESPONSES SYSTEMS.

EACH GIVEN BODY ORGAN AND BODY SYSTEM POSSESSES ITS ORGAN SPECIFIC CENTRAL INTELLIGENCE SYSTEM CENTERS (C. I. S.), AS WELL AS SYSTEM SPECIFIC CENTRAL INTELLIGENCE SYSTEM CENTERS (C. I. S.), WHICH ALL OF THESE INDEPENDENT ORGANS C.I. S. AND BODY SYSTEMS

C. I. S. ORDER THE TOTAL BODY POPULATION ELECTRONS AND NUCLEONS HOW THEY MUST PERFORM THE INTERNAL ELECTRON NUCLEON CHEMICAL, BIOLOGICAL AND PHYSICAL FUNCTIONS AND A. S. I. F.P. Mol. C.I.C. OF TOTAL BODY ORGANS AND SYSTEM.

THE TOTAL BODY ELECTRON NUCLEON POPULATION MEAN THE ALL ENTIRE ELECTRON-NUCLEON POPULATIONS THAT EXIST INSIDE THE BODY ORGANS AND BODY SYSTEMS OF A GIVEN ANIMAL SPECIES,

2 - EXOGENOUS PARTICLE CIRCULATION SYSTEMS (EX – PCS) FUNDAMENTAL PARTICLE CIRCULATION SYSTEMS BETWEEN TWO DIFFERENT ANIMALS CENTRAL INTELLIGENCE SYSTEMS ELECTRONS NUCLEONS

AIRBORN F.P. – I.I. – P. cl. EXCHAGE PHENOMENON BETWEEN TWO OR MORE ANIMALS CNS ELECTRONS NUCLEON POPULATIONS, THROUGH EX. PCS

THE AIRBORN EXOGENOUS PARTICLE CLOUD CIRCULATION SYSTEMS,

IN EXOGENOUS PARTICLE CIRCULATION SYSTEMS (EX – PCS) BETWEEN TWO OR MORE DIFFERENT INDIVIDUALS EXCHANGE THEIR CNS PRODUCED F.P. – I.I. – P. cl. BETWEEN EACH OTHER THROUGH AIRBORN PARTICLE CIRCULATION SYSTEMS, THE EXOGENOUS PARTICLE CLOUD CIRCULATION SYSTEMS (EX-PCS) IN AIR TRAVEL AT MULTIPLE DIRECTIONS

BETWEEN MULTI-INDIVIDUALS THROUGH CLOSED PARTICLE CLOUD CIRCULATION CYCLES,

EACH INDIVIDUAL'S CNS ELECTRONS NUCLEONS PRODUCE DIFFERENT PARTICLE CLOUDS, F.P. – I.I.- P.cl. AND EMIT PARTICLE CLOUDS AIRBORN, THE AIRBORN PARTICLE CLOUDS TRAVEL IN ALL DIRECTIONS IN AIR, THESE AIRBORN PARTICLE CLOUDS ARE CAPTURED BY DIFFERENT MULTIPLE INDIVIDUAL RECIPIENTS CNS ELECTRONS NUCLEONS, IN RECIPIENTS THE PARTICLE CLOUDS INTER INTO RECIPIENT INDIVIDUALS CNS ELECTRONS NUCLEONS, COMBINE WITH INTER ELECTRON NUCLEON PARTICLE COMPOUNDS, THESE PARTICLE CLOUD COMMPOUNDS AND EXOGENOUS PARTICLE CLOUDS INTERACT UNDER A. S. I. F.P. Mol. CIC BETWEEN DIFFERENT CNS CENTERS ELECTRONS NUCLEONS, FINALLY RECIPIENT ELECTRONS NUCLEONS UNDER INNORMOUS INTERACTIONS PRODUCE FUNDAMENTAL PARTICLE RESPONSE DIRECTLY RELATED TO DONNER'S INCOMING AIRBORN PARTICLE CLOUDS, THE PRODUCED RESPONSE PARTICLE CLOUDS EMIT INTO SPACE AND AIRBORN TRAVEL IN MULTI- AIR DIRECTIONS AND FINALLY CAPTURED BY RECIPIENTS CNS ELECTRONS NUCLEONS POPULATIONS FOR INTERACTIONS, THIS IS ONE CYCLE AIRBORN PARTICLE CLOUD CIRCULATIONS SYSTEMS WHICH CIRCULATE BETWEEN DIFFERENT RECIPIENTS AND PARTICLE CLOUD DONNER'S ELECTARONS NUCLEONS THROUGH PARTICLE CLOUD CIRCULATION SYSTEMS, THESE ARE PHENOMENONS OF AIRBORN MULTI- EXOGENOUS PARTICLE CLOUD CIRCULATIONS SYSTEM WHICH PRESENTLY CALLED COMMUNICATIONS.

3 - PHENOMENON OF EXOGENOUS PARTICLE CIRCULATION SYSTEMS ALTERNATIONS,

WITH INDOGENOUS PARTICLE CIRCULATION SYSTEMS ALTERNATIONS CIRCUITS SYSTEMS.

HOW INDIGENOUS F.P.- I.I.- P.cl. –GENESIS INTERACT WITH EXOGENOUS PARTICLE CLOUD – GENESIS?

HOW CNS ELECTRONS NUCLEONS RESPOND TO INCOMING PARTICLE CLOUDS OF OTHER INDIVIDUALS CNS ELECTRONS NUCLEONS PARTICLE CLOUDS?

THE PHENOMENON OF INTERACTIONS BETWEEN EXOGENOUS PARTICLE CLOUD CIRCULATION SYSTEMS (EX-PCS) PARTICLE CLOUDS, WHEN THEY ARE INTERACTING WITH INDIGENOUS PARTICLE CLOUD CIRCULATION SYSTEMS (I-PCS) BRIEFLY EXPLAINED IN PREVIOUS PAGES.

DURING INTER INDIVIDUAL COMMUNICATIONS BETWEEN TWO OR MORE INDIVIDUALS, THE DONNER'S CNS PRODUCED F.P. –I.I.- P. cl. UNDER A.S. I. F.P. Mol. C.I.C. EMIT FROM DONNER'S ELECTRONS NUCLEONS AND LEAVES OUT OF CNS ATOMS AS AIRBORN FREE PARTICLE CLOUDS IN SPACE, AND TRAVEL DIFFERENT DIRECTIONS.

THE RECIPIENT INDIVIDUALS CNS ELECTRONS NUCLEONS CAPTURE THESE AIRBORN FLOATING PARTICLE CLOUDS, AND ANALIZE THE F.P. – I.I. – P.cl. UNDER A. S. I. F.P. Mol. CIC IN RELATED DIFFERENT CNS CENTERS AND PRODUCE

RESPONSE – F.P. – I.I. – p. cl. OF RECIPIENT, AND RELEASE RESPONSE PARTICLE CLOUDS AIRBORN TO SPACE, UNDER SAME CYCLES OF PARTICLE CIRCULATION SYSTEMS THE DONNER CNS ELECTRONS NUCLEONS CAPTURE FLOATING AIRBORN PARTICLE CLOUDS OF RECIPIENTS, THESE PHENOMENONS CONTINUES BACK AND FORTH IN CLOSED CYCLES OF ALTERNATING INDIGENOUS PARTICLE CLOUDS CIRCULATION SYSTEMS ALTERNATE WITH EXOGENOUS PARTICLE CLOUD CIRCULATION SYSTEMS BACK AND FORTH, THIS IS BRIEFLY PHENOMENON OF COMMUNICATIONS BETWEEN DIFFERENT INDIVIDUALS THROUGH PARTICLE CLOUD EXCHANGE SYSTEMS.

ABOVE EXPLAINED PORCESS IS SUM OF ONE EXOGENOUS AIRBORN EXTERNAL BODY PARTICLE CLOUD CIRCULATION SYSTEMS (EX – PCS) CYCLE PHENOMENONS INTERACTIONS ALTERNATING WITH INDIGENOUS PARTICLE CLOUD CIRCULATION SYSTEMS, THIS IS PHENOMENON OF EX – PCS INTERACTIONS WITH I – PCS.

PIS OF BODY ORAGANS, AND PIS OF BODY SYSTEMS:

EACH GIVEN ANIMAL BODY ORGAN, POSSESSES INDEPENDANT CENTRAL ORGAN'S PARTICLE INTELLIGENCE SYSTEMS CENTER (O - PIS).

EACH GIVEN ANIMAL BODY SYSTEM, POSSESSES INDEPENDANT BODY SYSTEM'S CENTRAL PARTICLE INTELLIGENCE SYSTEM CENTERS (S- PIS).

ALL BODY ELECTRONS NUCLEON POPULATIONS FUNCTIONS, UNDER PARTICLE CLOUD ORDERS, THAT HAS BEEN ISSUED BY UPPER PARTICLE INTELLIGENCE SYSTEMS ORDERS, FUNCTIONS ACCORDINGLY, UNDER THE ISSUED ORDERS, ALSO RESPOND BACK TO THOSE INCOMING CENTRAL PARTICLE INTELLIGENCE SYSTEMS.

IN LIVING THINGS PARTICLE CLOUD CURRRENTS CIRCULATE BETWEEN DIFFERENT FUNDAMENTAL PARTICLE INTELLIGENCE SYSTEM CENTERS ALSO CONNECT ALL BODY INTELLIGENCE SYSTEMS AND ORGANS INTELLIGENCE SYSTEMS CENTERS WITH TOTAL ELECTRON NUCLEON POPULATIONS IN ANIMAL BODY, ALSO PARTICLE CIRCULATION SYSTEMS CARRY AND TRANSIT DIFFERENT NEEDED FUNDAMENTAL PARTICLES FROM PARTICLE DONNER LOCATIONS ELECTRONS NUCLEONS TO THE RECIPIENT INTERNAL ELECTRON NUCLEON SYSTEMS, AND REMOVE THE WASTE AND NOT NEEDED PATICLES OUT OF DIFFERENT NANO- LOCATIONS OF LIVING THINGS BODY, TWO OPPOSITE DIRECTION DIFFERENT AFFERENT AND EFFERENT PARTICLE CURRENTS AND FUNDAMENTAL PARTICLE CLOUD CIRCULATION SYSTEMS CONNECT ALL TOTAL BODY ELECTRON NUCLEON POPULATIONS TO EACH OTHER.

INDEPENDANT INTELLIGENCE SYSTEMS OF BODY ORGANS AND BODY SYSTEMS ELECTRONS AND NUCLEONS THROUGH INTER ELECTRON NUCLEON A. S. I. F.P. Mol. C.I.C. PRODUCE INSTRUCTIONS AND ORDER PARTICLE CLOUDS, AND THESE PARTICLE CLOUDS THROUGH NANO-NEURAL PARTICLE CLOUD CIRCULATION SYSTEMS PARTICLE CLOUD

CURRENTS, TRANSPORT TO ALL PERIPHERAL BODY ELECTRONS NUCLEONS POPULATIONS.

THE CENTRAL INTELLIGENCE SYSTEMS ISSUED PARTICLE CLOUDS ORDERS, INTER INTO BODY ORGANS PERIPHERAL ELECTRONS NUCLEONS SUBSYSTEM –UNITS, AND DIFFERENT BODY SYSTEMS AND ORGANS ELECTRONS NUCLEONS UNDER THOSE PARTICLE CLOUD ORDERS ACHIEVE PHYSIOLOGICAL, CHEMICAL AND BIOLOGICAL FUNCTIONS OF LIVING THINGS ACCORDING TO THE INSTRUCTIONS OF HIGHER INTELLIGENCE SYSTEM CENTERS ELECTRONS AND NUCLEONS PARTICLE CLOUD ORDERS.

THE TOTAL ELECTRON NUCLEON POPULATIONS OF LIVING THINGS BODY, THE TOTAL ELECTRON- NUCLEON POPULATION OF BRAIN, THE TOTAL ELECTRON NUCLEON POPULATIONS OF BODY ORGANS AND BODY SYSTEMS OF LIVING THINGS VARIES REMARKABLY FROM EACH OTHER.

PHENOMENON OF PSYCHE – GENESIS AND PHENOMENON OF THOUGHT CURRENT – GENESIS IN LIVING THINGS PHENOMENON OF S. Y. –F.P. – I.I. – P.cl. STORAGE INSIDE CNS NANO-UNITS (STORAGE OF KNOWLEDGE)

AIRBORN EXOGENOUS INCIDENT BIOFRIENDLY ELECTRIC FUNDAMENTAL PARTICLES, LIGHT FUNDAMENTAL PARTICLE INFORMATION IMAGE PARTICLE CLOUDS (Y. F.P. – I.I.- P. cl.), SONIC FUNDAMENTAL PARTICLE INFORMATION IMAGE PARTICLE CLOUDS (S. F.P. – I.I.- P.cl.) INTER INTO CNS INTERNAL ELECTRON NUCLEON CHEMICAL LAB.S AND COMBINE WITH INDOGENOUS PRE-EXISTING INTERNAL ELECTRON-NUCLEON PARTICLE –COMPOUNDS AND CONSTRUCT CNS ELECTRONS-NUCLEONS PARTICLE COMPOUNDS CONSTRUCTIONS UNDER REGENERATIVE A.S. I. F. P. Mol. C.I.C.

THIS IS PHENOMENON OF NEO- S. Y. – F.P. –I.I. – P.cl. _ COMPOUND CONSTRUCTION (NEO-GENESIS OF PARTICLE –CLOUD COMPOUNDS) OF THE CNS INTERNAL ELECTRONS NUCLEONS CONSTRUCTIONS, AS WELL AS THIS IS PHENOMENON OF STORING SONIC – PARTICLE INFORMATION –IMAGE PARTICLE CLOUDS, AND LIGHT FUNDAMENTAL PARTICLE INFORMATION IMAGE PARTICLE CLOUDS INSIDE THE CNS ELECTRONS AND NUCLEONS, IN THE CONSTRUCTION

FORMS OF LIGHT- SONIC PARTICLE CLOUD –COMPOUND CONSTRUCTIONS (S. Y. – F.P. – I.I.- P. cl. _ COMP. - GENESIS),

DURING LIFE, ANIMALS COMPILE STORAGE OF DIFFERENT ENVIRONMENTAL SUBJECTS PARTICLE –CLOUDS INFORMATIONS AND IMAGES (S. Y. – F.P. – I.I. – P.cl.) IN PARTICLE COMPOUND FORMS, ALL PARTICLE CLOUDS ARE STORED INSIDE CNS ELECTRONS –NUCLEONS, THESE INFORMATIONS AND IMAGES PARTICLE COMPOUND STORAGES INSIDE CNS ELECTRONS-NUCLEONS ARE STORAGE OF KNOWLEDGE AVAILABLE FOR RETRIEVAL TO OUT OF CNS ELECTRONS NUCLEONS ALL TIMES IN ANY MOMENTS.

THESE STORED PARTILE CLOUD COMPOUNDS ARE BUILDING BLOCKS OF KNOWLEDGE AND ARE USED IN LEARNING, EDUCATIONS, TRAININGS FOR INDIVDUALS, AND INTELLIGENCE CONSTRUCTIONS INSIDE CNS ELECTRONS NUCLEONS OF LIVING THINGS CNS AS WELL AS NON-LIVE ELECTRONS-NUCLEONS PARTICLE COMPOUND CONSTRUCTIONS, THESE STORED INFORMATIONS AND IMAGES ARE AVAILABLE FOR RETRIEVAL TO OUTSIDE PARTICLE-CLOUD –COMPOUND FORMS TO OUTSIDE ELECTRONS AND NUCLEONS.

THIS IS PHENOMENON OF PSYCHE-GENESIS AND STORAGE OF SCIENCE AND INFORMATION IMAGE INSIDE ELECTRONS NUCLEONS IN PARTICLE COMPOUND FORMS, INTERACTIONS OF THESE SONIC LIGHT FUNDAMENTAL PARTICLE COMPOUND INFORMATION IMAGE PARTICLE CLOUDS WITH EACH OTHER AS WELL AS WITH EXOGENOUS INCOMING S. Y. – F.P. – I.I. – P. cl. PRODUCE THOUGHT CURRENTS.

THE THOUGHT CURRENTS ARE SONIC LIGHT FUNDAMENTAL PARTICLE INFORMATION IMAGE PARTICLE CLOUD CURRENTS AND INTERACTIONS OF THESE PARTICLE CLOUD CURRENTS WITH EACH OTHER INSIDE CNS ELECTRONS NUCLEONS PRODUCE PSYCHE, BIOLOGICAL SENSING OF PARTICLE CLOUD CURRENTS SENSE AS THOUGHT CURRENTS, THIS IS PHENOMENON OF PSYCHE –GENESIS AND THOUGHT CURRENT-GENESIS.

THE ATTRACTION OF S.Y. –F.P. – I.I. – P. cl., AND OTHER PARTICLE CLOUDS FROM AIR

PHENOMENON OF PSYCHE-GENESIS AND THOUGHT CURRENT –GENESIS

THE ATTRACTIVE GRAVITON FORCES OF SENSORY ORGANS ELECTRONS-NUCLEONS ATTRACT S. Y. – F.P. – I.I. – P. cl. FROM AIR, PARTICLE CLOUDS INTER INTO SENSARY ORGANS ELECTRONS AND NUCLEONS, THE INCIDENT S. –F.P. – I.I. – P.cl., Y. - F.P.- I.I. – P.cl., T. – F.P. – I.I. – P. cl., ETC., THROUGH PARTICLE CLOUD NEURAL NANO –DUCTS PARTICLE CURRENT CIRCULATIONS SYSTEMS, INTER INTO DIFFERENT CNS SENSARY –CENTERS INTERNAL ELECTRONS-NUCLEONS SUBSYSTEM UNITS CHEMICAL LAB.S, AND COMBINE WITH INTER CNS INTERNAL ELECRTON NUCLEON PARTICLE COMPOUNDS, UNDER REGENERATIVE A. S. I. F.P. Mol. C.I.C., AND PRODUCE F.P. – I.I. – P. cl. _ COMPOUNDS, AND CONSTRUCT DIFFERENT CNS CENTERS NEW ELECTRONS NUCLEONS PARTICLE CLOUD MOLECULAR CONSTRUCTIONS,THIS PHENOMENON IS PARTICLE –CLOUD STORAGE PHENOMENON INSIDE CNS DIFFERENT SENSARY CENTERS ELECTRONS –NUCLEONS IN

THE FORMS OF PARTICLE CLOUD COMPOUNDS.

THE REVERSE OF ABOVE INTERACTIONS UNDER DEGENERATIVE A. S. I. F.P. Mol. C.I.C. CAUSE BREAK DOWN OF LARGER PARTICLE CLOUD COMBOUNDS INTO CONSTRUCTING SMALLER ORIGINAL PARTICLE CLOUD STRUCTURES, THESE AUTONOMOUS SEQUENTIAL INTERACTIONS AND PRODUCED SONIC- LIGHT PARTICLE CLOUD CURRENTS AND PARTICLES INTERACTIONS WITH EACH OTHER SENSE AS THOUGHT CURRENTS, PSYCHE AND THOUGHT CURRENTS SYSTEMS.

RELEASE OF LIGHT SONIC PARTICLE CLOUDS FREE FROM S. Y. – F.P. – I.I. – P. cl. _ COMPOUNDS FORMS FOLLOWING BREAK DOWN OF PARTICLE COMPOUNDS UNDER DEGENERATIVE A. S. I. F.P. Mol. C.I.C. PRODUCE FREE SONIC LIGHT PARTICLE CLOUDS CURRENTS, AND THIS PHENOMENON IS RETRIEVAL AND RELEASE OF PARTICLE CLOUDS TO OUT OF PARTICLE CLOUD COMPOUND FORMS AND PRODUCTIONS FREE PARTICLE CLOUD CURRENTS INSIDE CNS ELECTRONS NUCLEONS CIRCULATING BETWEEN DIFFERENT CNS CENTERS, WHICH IT IS THOUGHT CURRENT GENESIS AND PSYCHE GENESIS.

THE UNIVERSE UNDER THE GRAVITON LAWS AND ORDERS. THE GRAVITON FORCE IS THE PROPERTY OF THE MASSES.

THE CHEMICAL COMBINATION:

UNDER ATTRACTIVE GRAVITON FORCES TWO COMPATIBLE MASSES ATTRACT EACH OTHER AND COMBINE.

UNDER REPULSIVE GRAVITON FORCES ORDERS TWO NON – COMPATIBLE MASSES REJECT AND REFUSE COMBINATIONS.

THE ABOVE LAWS AND ORDERS START FROM PARTICLE SIZE MASSES, UP TO UNIVERSAL MASS SIZES WITH NO DIFFERENCE IN RULES.

UNDER ATTRACTIVE GRAVITON FORCE ORDERS REGENERATIVE AUTONOMOUS SEQUENTIAL CHEMICAL INTERACTIONS CHAINS AND CYCLES TAKE PLACE AND MASSES COMBINE TO EACH OTHER, BUT UNDER REPULSIVE GRAVITON ORDERS THE DEGENERATIVE AUTONOMOUS SEQUENTIAL CHEMICAL INTERACTIONS CYCLES CAUSE BREAK DOWN OF LARGE CONSTRUCTIONS, AND MASSES REFUSE COMBINATIONS,

THE ABOVE RULE APPLIES TO ALL OTHER AUTONOMOUS SEQUENTIAL CHEMICAL INTERACTION CYCLE BETWEEN ATOM-BASE OR NANO-UNIT-BASE OR CELL AND MICRO-UNIT BASE, AS WELL TO MACRO-UNIT AUTONOMOUS SEQUENTIAL CHEMICAL INTERACTION CHAINS WITH NO EXCEPTIONS,

THE GRAVITON CONTROL OF MASSES ALSO APPLY TO MICRO-UNIT STRUCTURES AS WELL. UNDER GRAVITON, ATTRACTIVE OR REPULSIVE FORCES ORDERS EITHER TWO MICRO-UNIT CHROMOSOMAL STRUCTURES OR TWO CELLS, THEY EITHER COMBINE OR REFUSE COMBINATIONS.

THE COMBINATION OR REPULSION OF LARGE MASSES SUCH AS PLANETS, GALAXIES, AS WELL AS UNIVERSES ARE ALSO ACHIEVE UNDER GRAVITON FORCES DIRECTLY OR EITHER COMBINED OR NOT, THAT IS THE ORDERS AND LAWS OF GRAVITON ACROSS THE UNIVERSE OVER THE MASSES.

UNIT OF GRAVITY

EXISTING GRAVITON FORCE QUANTITY IN ONE EARTH'S AIRBORN LIGHT FUNDAMENTAL PARTICLE, IS EQUALS TO: ONE UNIT GRAVITON.

RULES, LAWS AND ORDERS OF THE GRAVITON IN UNIVERSE

NUCLEUS GENESIS - PERIPHERON GENESIS PHENOMENONS, UNDER ORDERS OF GRAVITONS.

DEFINITION:

THE GRAVITON IS ENERGY FORCE IN MASS CONTENT, THE MASS WHO POSSESS MORE GRAVTON, IT IS POWERFUL AND DOMINANT, THROUGH GRAVITON ENERGY FORCES THE MASSES EITHER ATTRACT EACH OTHER AND COMBINE, OR UNDER REPULSIVE GRAVITON ENERGY FORCES MASSES REFUSE COMBINATIONS AND STAY APPART.

THE GRAVITON – UNIT: THE AMOUNT OF GRAVITON ENERGY IN ONE WHITE LIGHT PARTICLE IS EQUAL TO ONE UNIT GRAVITON FORCE.

1 -THE MOLECULES COMBINE TO EACH OTHER UNDER ATTRACTIVE GRAVITON FORCES AND PRODUCE MOLECULES.

2 - THE REPULSIVE GRAVITON FORCES PREVENT FROM COMBINATIONS, THROUGH MASSES REPULSIVE GRAVITON FORCES,

3 - GRAVITON ORDERS APPLY TO ALL MASS -UNITS SIZES, STARTING FROM FUNDAMENTAL PARTICLE MASS -UNITS, UP TO UNIVERSAL SIZE MASSES.

4 - UNDER GRAVITON FORCES THE NANO-UNITS, MICRO – UNITS AND MACRO-UNIT MASSES EITHER COMBINE UNDER ATTRACTIVE GRAVITON FORCES, OR STAY APART BY REPULSIVE GRAVITON ENERGY ORDERS.

5 - THE LESS WEIGHT MASS UNITS POSSESSES LESS GRAVITON

ENERGY FORCES, THE LESS WEIGHT MASS- UNITS HAVE LESS ENERGY CONTENTS OF ANOTHER KIND ENERGIES.

6 - THE HIGH WEIGHT MASS UNITS POSSESSES HIGH GRAVITON ENERGY FORCES AND HIGH ENERGY CONTENTS OF OTHER KIND ENERGIES RELATIVELY.

7 - THE HIGH WEIGHT MASS UNITS DOMINATE AND CONTROL THE LOW WEIGHT MASS –UNITS.

8 – THE DOMINANT MASS UNITS WITH HIGH GRAVTON ENERGY FORCES, MOSTLY TAKE CENTRAL LOCATION AND CONSTRUCT A NUCLEUS WITH HIGH ENERGY GRAVITON FORCES, THE DOMINANT MASSES THROUGH HIGH ATTRACTIVE GRAVITON FORCES FROM NUCLEUS ATTRACT RECESSIVE MASS UNITS IN THEIR PERIPHERY, AND COMBINE ALWAYS THE DOMINANT MASS UNITS STAY IN CENTER, AND RECESSIVE MASS - UNIT ALWAYS STAY IN PERIPHERONS.

9 – IN COMBINATION OF MASSES THE RECESSIVE LESS WEIGHT, LESS GRAVITON, LESS ENERGY NANO-UNIT STAY IN PERIPHERY AND CONSTRUCT PERIPHERON (THE ELECTRONS), BUT THE DOMINANT HIGH WEIGHT, HIGH ENERGY, HIGH GRAVITON NUCLEUS NANO-UNITS (THE NUCLEONS), ALWAYS BUILD A NUCLEUS, WHICH THE NUCLEUS ALWAYS TAKE CENTRAL POSITIONS AND FROM IT'S NUCLEAR LOCATION THROUGH HIGH GRAVITON ENERGY ATTRACT THE RECESSIVE LESS GRAVTON ENERGY ELECTRONS MASS UNITS.

10 – THE ABOVE RULES OF GRAVITON IS TRUE IN REGARD

TO FUNDAMENTAL PARTICLES ALSO, ALL OVER THE UNIVERSE IN ALL PLANETS ACROSS THE UNIVERSE.

11 - IN COMBINATION OF LESS MASS UNIT HYDROGEN, WITH HIGH MASS UNIT OXYGEN, THE DOMINANT GRAVITON FORCE OXYGEN TAKES CENTRAL NUCLEUS POSITION AND FROM THERE BY ATTRACTIVE GRAVITON ENERGY FORCES ATTRACT LESS WEIGHT, LESS ENERGY GRAVITON FORCE TWO HYDROGEN ATOMS, WHICH THE LESS WEIGHT HYDROGENS STAY IN PERIPHERY ORBITS, UNDER OXYGENS DOMINANT GRAVITON ORDERS AND CONSTRUCT PERIPHERON, AND OXYGEN STAY IN CENTER AND FROM NUCLEI POSITION CONTINUOUSLY CONTROL HYDROGEN.

12 – THE ABOVE LAWS AND RULES ORDERS OF GRAVITON IS TRUE IN REGARD TO MICRO – UNITS MASSES AS WELL, WHEN A DOMINANT HIGH WEIGHT, HIGH GRAVITON FORCE CELLS - NUCLEUS COMBINE WITH RECESSIVE LESS WEIGHT, LESS ENERGY GRAVITON FORCE CYTOPLASMA, ALWAYS THE CELL NUCLEUS TAKE CENTRAL POSITION UNDER DOMINANT HIGH FORCE GRAVITON ORDERS AND THE CYTOPLSMA WITH LESS GRAVITON FORCE ENERGY OBEY AND STAY AT PERIPHERY.

13 - ABOVE RULES ARE TRUE IN MACRO-UNITS MASSES AND UNIVERSAL SIZE MASS SYSTEMS, ALL RULES ARE THE SAME, THE LARGE CENTRAL UNIVERSE PLANETARY MASS SYSTEMS, FROM NUCLEI POSITION THROUGH GRAVITON ATTRACT AND CONTROL ALL OTHER PERIPHERAL SMALLR MULTI UNIVERSE MASS SYSTEMS, WHICH EACH ONE ARE

FIXED RELATIVELY IN CERTAIN POSITION.

MASS – UNIT

ONE GREEN COLOR LIGHT FUNDAMENTAL PARTICLE'S WEIGHT AT EARTH IS EQUAL TO ONE UNIT MASS

(ONE MASS UNIT IS EQUALS TO WEIGHT OF ONE Y. g. – F.P. AT EARTH LOCATION)

THE OTHER FUNDAMENTAL PARTICLES WEIGHTS QUANTIZED IN THE COMPARISON TO THE WEIGHT OF ONE GREEN LIGHT PARTICLE.

THE DIFFERENT FUNDAMENTAL PARTICLES (EVEN FROM THE SAME CLASS) IN OTHER PLANETS POSSESSES DIFFERENT WEIGHTS.

THE WEIGHTS, ENERGIES, GRAVITON FORCES OF ALL DIFFERENT CLASSES OF FUNDAMENTAL PARTICLES IN HARSH PLANETS, SUCH AS BLACK HOLES, ARE THOUSANDS FOLDS HIGHER THAN THE EARTH. AND THE CONSTRUCTED NANO-UNITS FROM CHEMICAL COMBINATIONS OF THESE FUNDAMENTAL PARTICLES PRODUCE ATOMS SUCH AS BLACK MATTER, DIFFERENT ELECTRONS, NUCLEONS CONSTRUCTIONS.

ENERGY UNITS

1 - ONE UNIT OF THERMAL ENERGY IS EQUAL TO THE HEAT ENERGY QUANTITY OF ONE INFRA- RED –PARTICLE ENERGY AMOUNT IN EARTH.

2 - ONE UNIT OF ELECTRIC –ENERGY IS EQUAL TO THE ELECTRIC ENERGY CONTENT OF ONE ULTRA VIOLET PARTICLE ENERGY AMOUNT ON EARTH.

3 - ONE UNIT OF LIGHT ENERGY IS EQUAL TO THE LIGHT ENERGY AMOUNT OF ONE GREEN COLOR PARTICLE ENERGY QUANTITY AT PLANET EARTH.

4 - CONSIDERING THE EARTH'S OTHER CLASSES OF FUNDAMENTAL PARTICLES, WHICH THEY HAVE ANOTHER KIND ENERGY CONTENTS, ALL ARE CONSIDERED TO BE ONE UNIT ENERGY QUANTITY. FOR EXAMPLE, THE GAMMA ENERGY UNIT, IS EQUAL TO ONE GAMMA PARTICLE ENERGY QUANTITY IN ONE GAMMA PARTICLE AT PLANET EARTH REGION. OR ONE X –ENERGY UNIT, IS EQUAL TO ONE X- PARTICLE'S ENERGY QUANTITY CONTENT, IN ONE X- FUNDAMENTAL PARTICLE CONTENT IN PLANET EARTH REGION, ETC.

THERE ARE LARGE NUMBERS OF OTHER PLANETS AND GALAXIES WHICH THEIR FUNDAMENTAL PARTICLES ENERGY – CONTENTS ALL ARE BIO-LETHAL AND POWERFUL ELECTRIC, THERMAL, LIGHT, SONIC, GAMMA, X. FUNDAMENTAL PARTICLES WHICH THEIR ENERGY- CONTENTS AND ENERGY INTENSITIES ARE THOUSANDS TO MILLIONS FOLDS HIGHER THAN THE EARTH'S PARTICLES.

FUNDAMENTAL-PARTICLE COLONY

FUNDAMENTAL PARTICLES TEND TO COLONIZE AND TRAVEL IN WAVE SHAPES. THE TOTAL FUNDAMENTAL PARTICLE-POPULATION NUMBERS IN EACH GIVEN ONE WAVE-LENGTH IS ONE FUNDAMENTAL PARTICLE-COLONY. FOR EACH GIVEN FUNDAMENTAL PARTICLE THE WAVE LENGTH, FREQUENCY, FUNDAMENTAL PARTICLE COLONY AND TOTAL FUNDAMENTAL PARTICLE POPULATION NUMBER PER COLONY ARE WELL DEFINED, CONSTANT RELATIVELY FIXED, CERTAIN FOR ANY GIVEN FUNDAMENTAL PARTICLE.

THE TOTAL PARTICLE -POPULATION NUMBER AT EACH GIVEN WAVE –LENGTH IS EQUAL TO ONE FUNDAMENTAL-PARTICLE COLONY, PARTICLE COLONY FOR ANY GIVEN PARTICLE ALWAYS IS FUNDAMENTAL PARTICLES SPECIFIC AND DIFFERENT FUNDAMENTAL PARTICLE CLASSES POSSESSES DIFFERENT FUNDAMENTAL PARTICLES COLONIES.

HOMOGENEOUS PARTICLE COLONIES THE TOTAL PARTICLE-POPULATION AT ONE WAVE –LENGTH AND ARE ALL COMPOSED FROM ONE KIND OF FUNDAMENTAL PARTICLE CLASS HOMOGENOUSLY,

AT HETEROGENEOUS FUNDAMENTAL PARTICLE COLONIES, THERE ARE DIFFERENT KINDS OF FUNDAMENTAL PARTICLES IN ONE WAVE LENGTH COLONY HETEROGENEOUSLY CONSTRUCTING PARTICLE COLONIES. IN HETEROGENEOUS PARTICLE COLONIES, THERE ARE MANY KIND DIFFERENT FUNDAMENTAL PARTICLES FROM DIFFERENT CLASSES IN ONE WAVE LENGTH COLONY.

MIXED NANO CLOUDS, FLOATING MASSES

IN SPACE AND THE AIR AROUND PLANET EARTH ARE MIXED CLOUDS, COMPOSED FROM ATOM CLOUDS, FUNDAMENTAL PARTICLE INFORMATION IMAGE PARTICLE CLOUDS (F.P. – I.I. – P. cl.), FUNDAMENTAL PARTICLE CLOUDS, NANO UNIT CLOUDS, NANO UNITS INFORMATION IMAGE CLOUDS (N. U. – I.I. – Cl.), ATOM BASED INFORMATION IMAGE CLOUDS (A. – I.I. – cl.), ATOM CLOUDS, MICRO UNIT CLOUDS, MICRO UNIT INFORMATION IMAGE CLOUDS, FLOATING DIFFERENT SIZES MACRO MASSES IN SPECE, UP TO PLANETARY OR LARGER SIZE MASSES ACROSS UNIVERSE, ETC.

FUNDAMENTAL PARTICLE CLOUDS (F. P. – P.cl.)
FUNDAMENTAL PARTICLE INFORMATION
– IMAGE PARTICLE – CLOUDS
(F.P. – I. I. –P. cl.)

THE MOST COMMON AIRBORN PARTICLES OF EARTH'S ATMOSPHERE, ARE BIOFRIENDLY EARTH'S FUNDAMENTAL PARTICLES CLOUDS (F.P. – I.I.- P. cl.), LIGHT- FUNDAMENTAL

PARTICLE CLOUDS (Y. - F.P. cl.), SONIC FUNDAMENTAL PARTICLE CLOUDS (S. – F.P. cl.), BIOFRIENDLY THERMAL FUNDAMENTAL PARTICLE CLOUDS (T. – F.P. cl.), AND BIO-FRIENDLY AIRBORN ELECTRIC FUNDAMENTAL PARTICLE CLOUDS (E. - F.P. cl.). THERE ARE ALSO LARGE NUMBERS OF UNDISCOVERED, NON SENSIBLE, BIOFRIENDLY FUNDAMENTAL PARTICLES IN AIR OF PLANET EARTH, AND THESE PARTICLES CONSTRUCT MOST OF THE PLANET EARTHS ELECTRON- NUCLEON PARTICLE COMPOUND CONSTRUCTIONS. THE FREE AIRBORN BIO-LETHAL FUNDAMENTAL PARTICLE CLOUDS SUCH AS, X- PARTICLES, GAMMA PARTICLES ARE RARE IN EARTH'S ATMOSPHERE. THE OTHER PLANETS PARTICLE CLOUDS MOSTLY ARE BIOHOSTILE, AND ARE NOT LIVABLE.

THE ATOM - CLOUDS, ATOM INFORMATION-IMAGE CLOUD, NANO – UNIT INFORMATION IMAGE - CLOUD (N. U. - I. I. _ cl.), MICRO –UNITS, CELLS, AND MICRO-UNIT CLOUDS, ETC., ARE ANOTHER BIO-FRIENDLY OR BIOHOSTILE STRUCTURES OF PLANET EARTHS ATMOSPHERE CONTENTS,

PHENOMENON OF FUNDAMENTAL PARTICLE CLOUDS

FUNDAMENTAL PARTICLE INFORMATION IMAGE PARTICLE CLOUD –GENESIS (F.P.- I.I. – P. cl. – GENESIS) PHENOMENON.

DEFINITION OF PARTICLE CLOUDS

THE FUNDAMENTAL PARTICLES CONSTRUCT PARTICLE COPIES FROM ENVIRONMENTAL SUBJECTS, IN WHICH THESE

COPIES ARE EXACTLY SIMILAR TO THE ORIGINAL SUBJECTS, AND IT IS A VIRTUAL PARTICLE COPY FROM ORIGINAL SUBJECTS.

"THE FUNDAMENTAL PARTICLE COPY MADE, FROM SUBJECTS, ARE SUBJECT'S PARTICLE –CLOUDS"

THE F.P. – CMVS IN PLANET EARTH MOSTLY ARE CONSTRUCTED FROM INDIGENOUS BIOFRIENDLY FUNDAMENTAL PARTICLE INFORMATION IMAGE PARTICLE –CLOUDS (F.P. – I.I. – P. cl.) OF LIGHT PARTICLES, SONIC PARTICLES, ELECTRIC PARTICLES, THERMAL FUNDAMENTAL PARTICLES INFORMATIONS AND IMAGES PARTICLE CLOUDS, WHICH ARE INDIGENOUS EARTH FUNDAMENTAL PARTICLES.

THIS IS THE FUNDAMENTAL PARTICLE INFORMATION IMAGE PARTICLE CLOUD GENESIS (F.P. – I.I. – P.cl. –GENESIS) PHENOMENON IN EARTH.

THE PARTICLE CLOUD-GENESIS IS A VIRTUAL REALITY – COPY –GENESIS PHENOMENON FROM EARTH'S SUBJECTS AND EXISTING THINGS FROM START AND INCEPTION OF PLANET EARTH, UP TO PRESENT TIME, AND WILL CONTINUE IN FUTURE AS WELL,

THE UNIVERSAL EXISTING SUBJECTS, AND EXISTING THINGS, VIRTUAL REALITIES COPIES ARE CONSTRUCTED FROM DIFFERENT KIND ANOTHER TYPES FUNDAMENTAL PARTICLES

CLOUDS WHO ARE INDIGENOUS FOR THOSE GIVEN LOCATION PLANETS, IN DIFFERENT UNIVERSAL LOCATIONS THE VIRTUAL REALITY COPIES ARE DIFFERENT PARTICLE CLOUDS CONSTRUCTED TYPES, SUCH AS X-PARTICLE CLOUDS SUBJECT COPIES, GAMMA-PARTICLE CLOUDS VIRTUAL REALITY COPIES, ETC.

AT EARTH THE MOST COMMON FUNDAMENTAL PARTICLE-CLOUDS ARE: Y. - F.P. – I.I.- P.cl., S.- F.P. I.I.- P.cl., T. F.P.- I.I. – P.cl., E. F.P. – I.I. – P.cl.

IN DIFFERENT PLANETS INCLUDING THE PLANET EARTH, THE INDIGENOUS ATOMS ALSO CONSTRUCT AND PRODUCE NANO-UNIT -CLOUDS FROM OUTSIDE EXISTING THINGS AND SUBJECTS, ATOM INFORMATION IMAGE CLOUDS ALSO ARE ENVIRONMENTAL SUBJECTS COPIES MADE BY ATOM – CLOUDS.

FUNDAMENTAL PARTICLE INFORMATION IMAGE PARTICLE CLOUD COMPOUNDS GENESIS INSIDE ELECTRONS NUCLEONS:

PHENOMENON OF F.P. – I.I. – P. cl. _ COMP. – GENESIS, AND PARTICLE CLOUD STORAGE INSIDE ELECTRONS NUCLEONS.

1 – FUNDAMENTAL PARTICLE INFORMATION IMAGE PARTICLE CLOUD STORAGE INSIDE ELECTRONS NUCLEONS:

THE DIFFERENT BIOFRIENDLY FUNDAMENTAL PARTICLE CLOUDS, SUCH AS SONIC PARTICLE CLOUDS, LIGHT PARTILE CLOUDS, ELECTRIC PARTICLE CLOUDS, THERMAL PARTICLE CLOUDS, ETC., AIRBORN INTER INTO DIFFERENT ELECTRON – NUCLEONS SUBSYSTEMS UNITS CHEMICAL LAB.S, AND THROUGH A. S. I. F. P. Mol. C.I.C. COMBINE WITH INTER ELECTRON - NUCLEON PARTICLE COMPOUNDS, AND PRODUCE NEW INTERNAL ELECTRON NUCLEON FUNDAMENTAL PARTICLE INFORMATION IMAGE PARTICLE CLOUD – COMPOUND CONSTRUCTIONS, CONSTRUCT NEW MOLECULAR STRUCTURE ELECTRON NUCLEONS UNDER REGENERATIVE A. S. I. N.-U. C. I. C.

THIS PHENOMENON IS NEO- PARTICLE COMPOUND GENESIS, AS WELL AS IT CONSTRUCT NEW ELECTRON NUCLEON CONSTRUCTIONS, AND THIS IS PHENOMENON OF NEO - ELECTRONS NUCLEONS GENESIS, THROUGH CONSTRUCTIONS OF NEW FUNDAMENTAL PARTICLE INFORMATION IMAGE PARTICLE CLOUD COMPOUNDS (F.P. – I.I. – P.cl. _ COMPOUNDS) INSIDE ELECTRONS NUCLEONS.

THE CNS ELECTRONS NUCLEONS STORE OUT SIDE ENVIRONMENTAL INFORMATION IMAGE PARTICLE CLOUDS IN FORMS OF PARTICLE CLOUD- COMPOUNDS INSIDE CNS ELECTRON NUCLEON CONSTRUCTIONS, THROUGH INFORMATION IMAGE PARTICLE CLOUD STORAGE IN FORMS OF PARTICLE CLOUDS COMPOUNDS INSIDE CNS ELECTRONS NUCLEONS, THIS PHENOMENON IS STORING AND RECORDING OF OUTSIDE ENVIRONMENTAL KNOWLEDGE INSIDE ELECTRONS NUCLEONS IN THE FORMS OF F.P. – I.I.- P. cl. _COMPOUND MOLECULAR CONSTRUCTIONS OF CNS ELECTRONS NUCLEONS, THIS IS PHENOMENON OF STORING SCIENCE AND KNOWLEDGE IN BRAIN.

THE PHENOMENON OF STORING INFORMATION IMAGE PARTICLE CLOUDS INSIDE NON LIVE ELECTRONS NUCLEONS ACROSS UNIVERSE IN ALL PLANETS TAKE PLACE SIMILARLY THROUGH USE OF INDIGENOUS EXISTING PARTICLES INTERACTIONS, THIS IS PHENOMENON OF UNIVERSAL EVENTS INFORMATION IMAGE STORAGE INSIDE ELECTRONS NUCLEONS AS AN UNIVERSAL ENCYCLOPEDIA, THIS PHENOMENON IS GOING ON SINCE INCEPTION OF UNIVERSE UP TO PRESENT TIMES, AND IT WILL CONTINUE IN FUTURE AS

WELL TILL INDEFINITE.

THE MEMORIZATION PHENOMENON
2 - FUNDAMENTAL PARTICLE INFORMATION
IMAGE PARTICLE CLOUDS (F.P. – I.I. – P.
cl.) RETRIEVAL PHENOMENON.

PHENOMENON OF PSYCHE GENESIS, THOUGHT CURRENT GENESIS, AND REMEMBERING:

BREAKING DOWN OF LARGE MOLECULAR PARTICLE CLOUD STRUCTURES SUCH AS F.P. – I.I. – P. cl. _ COMPOUNDS INSIDE CNS ELECTRONS NUCLEONS, TAKE PLACE UNDER DEGENERATIVE A. S. I. F.P. Mol. C.I.C., AND UNDER THIS PROCESS THE COMBINED INFORMATION-IMAGE PARTICLE CLOUDS COMPOUNDS BREAK DOWN INTO CONSTRUCTING SMALLER MOLECULAR STRUCTURES, AND CAUSE RELEASE OF (F.P. – I.I.- P.cl.) PARTICLE CLOUDS, AS FREE INTERACTING PARTICLE – CLOUDS CURRENTS INSIDE THE CNS ATOMS PARTICLE CLOUD CIRCULATION SYSTEMS.

UNDER DEGENERATIVE A. S. I. F.P. – P.cl. Mol. C.I.C., WHICH IT IS REVERSE DIRECTIONS OF ORIGINAL REGENERATIVE AUTONOMOUS SEQUENTIAL CHEMICAL INTERACTIONS CYCLES, CAUSE BREAK DOWN OF LARGE PARTICLE CLOUD COMPOUNDS INTO SMALLER FREE MOLECULES, AND THIS PHENOMENON IS PARTICLE CLOUD RETRIEVAL PROCESS, AND CAUSE RELEASE OF ENVIRONMENTAL SUBJECTS F.P.- I. I. - P. cl. FREE, AND THESE PARTICLE CLOUDS INTER INTO CNS PARTICLE CIRCULATION SYSTEMS AS FREE INTERACTING F.P. –I.I. - P. cl. (FREED KNOWLEDGE OF EVENTS, IN PARTICLE

CLOUD FORMS), THIS IS RETRIEVAL PHENOMENON OF PARTICLE CLOUDS INTO FREE PARTICLE CLOUDS FORMS, WHICH UNDER COMPUTER TECHNOLOGY CAN BE WATCHED ON TV SCREENS.

THE FREE INFORMATION IMAGE PARTICLE CLOUD CURRENTS (F.P. – I.I.- P. cl. CURRENTS) THROUGH PARTICLE CIRCULATION SYSTEMS TRAVEL BETWEEN DIFFERENT BRAIN CENTERS AND INTERACT WITH PARTICLE CLOUDS OF DIFFERENT CNS CENTERS ELECTRONS NUCLEONS, THESE PARTICLE CLOUDS CURRENTS AND INFORMATION IMAGE INTERACTIONS WITH ALL CNS ELECTRONS NUCLEONS PARTICLE COMPOUNDS, BIOLOGICALLY SENSE AS THOUGHT CURRENTS AND PSYCHE, THIS IS PHENOMENON OF THOUGHT CURRENT GENESIS AND PHENOMENON OF PSYCHE GENESIS.

IN THE PHENOMENON OF RE- MEMORIZATION OF PAST EVENTS, WHEN THE PAST EVENTS RECORD IN F.P. – I. I. – P.cl. COMPOUND FORMS, INSIDE THE CNS ELECTRONS NUCLEONS, UNDER RETRIEVAL PROCESS, WHEN LARGE MOLECULAR INFORMATION IMAGE PARTICLE CLOUD COMPOUNDS BREAK DOWN INTO SMALLER MOLECULAR STRUCTURES UNDER DEGENRATIVE AUTONOMOUS CHEMICAL INATERACTIONS ONE BY ONE.

DURING RETRIEVAL PHASE, THE RELEASED PARTICLE INFORMATION AND IMAGE CLOUDS ARE EXACT VIRTUAL COPIES OF THE ORIGINAL SUBJECTS, AND SCENES, DURING RETRIEVAL OF PARTICLE CLOUDS THE EVENT OF THE

PAST RE-APPEARING AGAIN INSIDE CNS ELECTRONS NUCLEONS EXACTLY THE SAME AS ORIGINAL SUBJECTS PLAYING SCENES AND EVENTS SIMILARLY, THIS RETRIEVAL PHENOMENON AS KNOWN AS CNS MEMORIZATION AND REMEMBERING PHENOMENON, IN REALITY THE VIRTUAL PARTICLE CLOUDS INSIDE CNS ELECTRONS NUCLEONS, LOOKS EXACTLY THE SAME AS ORIGINAL SCENES- SUBJECTS AND EVENTS, AND BOTH ORIGINAL SUBJECTS AND VITUAL CLOUDS ARE IDENTICAL, LOOKS SIMILAR, THAT IS THE REASONS THE THOUGHTS,PSYCHE, THOUGHT CURRENTS WHICH ARE PARTICLE CLOUD INTERACTIONS AND CURRENTS INSIDE THE CNS ELECTRONS NUCLEONS, ALL LOOKS TWINS EVENTS OF THE ORIGINAL SUBJECTS AND SCENES.

INTERNAL ATOM, INTER-ELECTRON, INTER-NUCLEON, PARTICLE CIRCULATION SYSTEMS (PCS):

THE ONGOING PARTICLE CURRENTS AND PARTICLE CLOUDS CIRCULATIONS SYSTEMS IN ATOM'S PERIPHERON, NUCLEUS, AND PERIPHERAL ATOM SPACE (PAS) ARE AS FOLLOWING:

1 - THE PERIPHERON'S PARTICLE CIRCULATION SYSTEMS (P.P.C.S.):

THE PERIPHERON HAVE TWO PARALLEL PARTICLE CIRCULATION SYSTEMS AND PARTICLE CURRENTS,

THE FIRST PARTICLE CURRENTS CIRCULATE FROM PERIPHERAL ATOM SPACE TOWARD THE ATOM'S CENTER, THIS CENTRIPETAL P.P.C.S. BRINGS NEEDED NECESSARY PARTICLE CLOUDS AND NEEDED NANO-UNITS, PARTICLES FROM OUTSIDE ATOM AND PERIPHERAL ATOM SPACE INTO ATOM, TO BE USED FOR NEEDED NECESSARY CHEMICAL, PHYSICAL, BIOLOGICAL FUNCTIONS, A.S. I. F.P. Mol. C.I.C. AND PARTICLE CONSTRUCTIONS OF THE ELECTRONS, AND NEEDED PARTICLES AND PARTICLE CLOUDS TO ACHIEVE INTERNAL ELECTRON NANO-TASKS,

THE CENTRIPETAL P.P.C.S. BRING NEEDED PARTICLES AND PARTICLE CLOUDS FROM OUTSIDE FOR CONSTUCTIONS OF INTERNAL ELECTRON NUCLEON PARTICLE COMPOUND COSTRUCTIONS, AND CONSTRUCT ELECTRONS NUCLEONS THROUGH USE OF THESE INCOMING PARTICLES AND PARTICLE CLOUDS,

THE SECOND PPCS, RUNS AND CIRCULATE CENTRIFUGALLY PARALLEL BUT IN OPPOSITE DIRECTIONS OF THE FIRST PARTICLE CIRCULATION SYSTEM, THE SECOND PPCS CURRENTS FLOW AND CIRCULATE FROM ATOM'S CENTER OR NUCLEUS TOWARD THE PERIPHERAL ATOM IN CENTRIFUGAL DIRECTION PARTICLE CLOUD CURRENTS AND PARTICLE CIRCULATION SYSTEMS,

THE SECOND PPCS TRANSPORT AND CARRY OUT UN- NEEDED PARTICLE BY PRODUCTS, ALSO CARRY OUT CONSTRUCTED AND PRODUCED PARTICLE CLOUDS AND PARTICLE COMPOUNDS WHICH HAVE BEEN PRODUCED BY ELECTRONS NUCLEONS FOR THE USE OF ANOTHER EXTERNAL NANO-UNITS NEEDS, FOR USE OF ANOTHER AT RECIPIENT NANO-UNITS CONSTRUCTIONS.

THE PERIPHERONS PPCS COMPOSED FROM TWO DISTINCT CIRCULATION SYSTEMS, ONE PPCS CIRCULATE IN PERIPHERON, AND THE OTHER PPCS IS INTERNAL ELECTRONS PARTICLE CIRCULATION SYSTEMS,

THE PERIPHERON'S PARTICLE CIRCULATION CURRENTS ALSO CROSS PERIPHERON LINES INTO NUCLEUS AND JOIN

TO NUCLEI PARTICLE CIRCULATION SYSTEMS, THE PERIPHERON PARTICLE CIRCULATION SYSTEMS FURTHER MORE IN PERIPHERON BRANCH INTO DIFFERENT SMALL SIZE INTERNAL ELECTRONS BRANCHES AND CONNECT TO INTERNAL ELECTRONS PARTICLE CIRCULATION SYSTEMS,

2 - THE NUCLEAR PARTICLE CIRCULATION SYSTEMS (N.P.C.S.), AND PARTICLE CLOUD CURRENTS:

THE NUCLEUS PARTICLE CIRCULATION SYSTEMS INCLUDE, INTERNAL PROTONS CIRCULATION SYSTEM, INTERNAL NEUTRONS PARTICLE CIRCULATION SYSTEMS, AND INTERNAL NUCLEUS PARTICLE CIRCULATION SYSTEMS, THE INTERNAL NUCEUS PARTICLE CIRCULATION SYSTEMS PARTICLE CURRENTS SIMILAR TO P.P.C.S., FLOWS AT TWO OPPOSITE DIRECTIONS PARTICLE CIRCULATION SYSTEMS CURRENTS.

THE FIRST NUCLEAR PARTICLE CIRCULATION SYSTEMS (N.P.C.S.) CURRENTS FLOW CENTRIPETALLY FROM PERIPHERON TOWARD THE NUCLEAR CENTER DIRECTIONS, AND BRINGS NEEDED PARTICLES AND PARTICLE CLOUDS INTO NUCLEUS, PROTONS, NEUTRONS, FOR NEUTRON-GENESIS AND PROTON-GENESIS AND CONSTRUCTIONS OF THE NUCLEON PARTICLE COMPOUND CONSTRUCTIONS, THE INCOMING PARTICLES AND PARTICLE CLOUDS ARE USED TO CONSTRUCT INTERNAL NUCLEONS PARTICLE COMPOUNDS AS WELL ARE USED ACHIEVE NUCLEONS PHYSICAL CHEMICAL BIOLOGICAL NANO-TASKS.

THE SECOND N. P.C.S. AND PARTICLE CURRENTS FLOW

PARALLEL TO THE FIRST NUCLEAR PARTICLE CIRCULATION SYSTEMS CENTRIFUGALLY, AND CARRY OUT UN-NEEDED PARTICLE BY PRODUCTS, AND PRODUCED PARTICLE MOLECULES FOR USE OF NANO-UNIT PARTICLE COMPOUND CONSTRUCTIONS AND FOR USE OF OTHER NANO-TASKS.

3 – THE PARTICLE CIRCULATION SYSTEMS AND PARTICLE CURRENTS OF PERIPHERAL ATOM SPACE (PAS):

THE PERIPHERAL ATOM SPACE PARTICLE CIRCULATION SYSTEM'S CURRENTS, CIRCULATE BETWEEN DIFFERENT ATOMS AND IN THE PERIPHERY OF ATOMS, THE ATTRACTIVE GRAVITON FORCES OF ELECTRONS, NUCELONS ATTRACT THEIR NEEDED PARTICLES FROM THEIR SPACE (PAS), THE PARTICLE CLOUDS AND INFORMATION IMAGE PARTICLE CLOUDS FROM PERIPHERAL ATOM SPACE (P.A.S.) UNDER ATTRACTION GRAVITON FORCES INTER INTO PERIPHERON PARTICLE CIRCULATION SYSTEMS (PPCS), AND THROUGH THE PERIPHERONS PARTICLE CIRCULATION SYSTEMS (PPCS) THE ELECTRONS NEEDED PARTICLES, PARTICLE CLOUDS CIRCULATE INTO DIFFERENT ELECTRONS IN THE PERIPHERON, THEREAFTER THROUGH NUCLEAR PARTICLE CIRCULATION SYSTEMS (N.P.C.S.) THE INCOMING PARTICLES , F.P.- I.I.- P. cl., AND OTHER PARTICLE CLOUDS CIRCULATE INSIDE NUCLEUS AND BRANCH TO DIFFERENT EXISTING PROTONS AND NEUTRONS SUBSYSTEM –UNITS CONSTRUCTIONS, AND COMBINE WITH THEIR PARTICLE COMPOUNDS OR USED FOR ACHIEVING DIFFERENT NANO- FUNCTIONS INSIDE THE DIFFERENT ELECTRONS NUCLEONS.

THE PPCS, AND NPCS PROVIDE NEEDED PARTICLE CLOUDS, FUNDAMENTAL PARTICLES, F.P. – I. I. - P.cl. INTO DIFFERENT ELECTRONS AND NUCLEONS SUBSYSTEM - UNITS CHEMICAL LAB.S, THERE DIFFERENT ELECTRONS NUCLEONS PARTICLE COMPOUNDS UNDER A. S. I. F.P. Mol. C.I.C. COMBINE WITH INCOMING PARTICLE CLOUDS AND PRODUCE NEW CONSTRUCTIONS PARTICLE COMPOUNDS, WHICH THESE ARE THE PARTICLE COMPOUNDS CONSTRUCTING THE DIFFERENT ELECTRONS NUCLEONS PARTICLE STRUCTURES, THID IS PROCESS OF ELECTRONS NEO-GENESIS AND NUCLEONS NEO-GENESIS PHENOMENONS.

NANO-UNITS

ELECTRONS, POSITRONS, PROTONS, BARYONS, NEUTRONS, QUARKS, SUBSYSTEMS NANO – UNITS, LEPTONS, ELECTRON SYSTEMS, NUCLEON SYSTEMS AND SUBSYSTEMS NANO-UNITS, MESONS, ATOMS, ATOM-BASE MOLECULES, FUNDAMENTAL BASE MOLECULES, PLUS MANY OTHER KNOWN OR NON-KNOWN QUANTUM SIZE STRUCTURES ARE ALL EXAMPLES OF NANO-UNITS. PRESENTLY MOST NANO-UNITS ARE NO DISCOVERED AND NOT KNOWN YET, AND THE ABOVE ARE SOME EXAMPLES OF NANO-UNITS.

ALL NANO-UNIT ARE INDEPENDENT QUANTUM SIZE STRUCTURES. FUNDAMENTAL PARTICLE CONSTRUCTIONS OF NANO –UNIT ALL ARE WELL- DEFINED FIXED NANO-UNIT SPECIFIC PARTICLE –COMPOUND CONSTRUCTIONS. ALL NANO-UNITS POSSESS NANO-UNIT SPECIFIC PHYSICAL CHEMICAL BIOLOGICAL PROPERTIES.

CONSTRUCTION OF NANO –UNITS

MOSTLY DIFFERENT KIND FUNDAMENTAL PARTICLES SUCH AS ELECTRIC PARTICLE, LIGHT PARTICLES, SONIC PARTICLES,

THERMAL PARTICLES, ETC. JOINTLY CONSTRUCT DIFFERENT NANO-UNITS, ELECTRONS. NUCLEONS PARTICLE COMPOUND CONSTRUCTIONS, STILL THERE ARE MANY ELECTRONS, NUCLEONS, AND ATOMS WHOSE PARTICLE COMPOUNDS HAVE BEEN CONSTRUCTED FROM FEW SEVERAL DIFFERENT KIND FUNDAMENTAL PARTICLE COMPOUND CONSTRUCTED ELECTRON NUCLEON CONSTRUCTIONS, SUCH AS ELECTRIC PARTICLE ATOMS, ELECTRONS, NUCLEONS, OR THERMAL PARTICLE COMPOUND ELECTRONS AND NUCLEONS, OR SONIC PARTICLE ATOMS, AND LIGHT PARTICLE COMPOUND CONSTRUCTED ELECTRONS NUCLEONS AND ATOMS, ETC.

ELECTRONS NUCLEONS FUNDAMENTAL
PARTICLE CONSTRUCTIONS
THE FUNDAMENTAL PARTICLES CONSTRUCTIONS OF
ELECTRONS, NUCLEONS, ATOMS, NANO-UNITS

THE SPECIFIC FUNDAMENTAL PARTICLE CONSTRUCTED
ELECTRONS, NUCLEONS, ATOMS, NANO-UNITS

MOST OF EARTH'S ELECTRONS NUCLEONS PARTICLE COMPOUNDS CONSTRUCTED FROM NURMEROUS DIFFERENT KIND BIOFIENDLY FUNDAMENTAL PARTICLES, THE FOLLOWING FUNDAMENTAL PARTICLE COMPOUNDS SUCH AS: THE ELECTRIC PARTICLE COMPOUNDS, THE LIGHT PARTICLE COMPOUNDS, SONIC PARTICLE COMPOUNDS, THE THERMAL PARTICLE COMPOUNDS, OCCASIONALLY X-PARTICLES AND GAMMA PARTICLES COMPOUNDS, ETC., ARE THE

MOST COMMONLY USED PARTICLES COMPOUNDS THAT CONSTRUCTING THE EARTHS ELECTRONS NUCLEONS PARTICLE COMPOUND AT EARTH, CONTRARY TO OTHER PLANETS WHICH THEIR PARTICLE CONSTRUCTED MOLECULAR COMPOUNDS MOSTLY ARE BIOHOSTILE.

ALSO THERE ARE MANY ELECTRONS NUCLEONS, THAT THEIR FUNDAMENTAL PARTICLE COMPOUNDS MOLECULAR COSTRUCTIONS ARE MADE FROM A FEW KNOWN SPECIFIC FUNDAMENTAL PARTICLES, SUCH AS H-O-H, OR CRYSTALINE'S ELECTRONS NUCLEONS, WHICH THE H- O –H AND CRYSTALINES DOMINENT CONSTRUCTING FUNDAMENTAL PARTICLES ONLY ARE DIFFERENT CLASS LIGHT FUNDAMENTAL PARTICES, THESE ATOMS ELECTRONS NUCLEONS HAVE BEEN CONSTRUCTED FROM MANY DIFFERENT KIND LIGHT PARTICLE COMPOUNDS, SUCH AS RED COLOR LIGHT PARTICLE COMPOUNDS, BLUE OR VIOLET PARTICLE COMPOUNDS, YELLOW OR GREEN PARTICLE COMPOUNDS, ETC.

WHEN THE H- O –H, OR CRYSTALINE ELECTRONS NUCLEONS LIGHT- PARTICLE _ COMPOUNDS COMBINE WITH DIFFERENT TYPES OTHER AIRBORN LIGHT –PARTICLES, X - FUNDAMENTAL PARTICLES, GAMMA PARTICLES, OR ELECTRIC PARTICLES, ETC., THROUGH A. S. I. F.P. Mol. C.I.C., THE CHEMICAL COMBINATIONS PARTICLES BY PRODUCTS ARE RELEASE OF DIFFERENT KIND AND COLOR FREED LIGHT PARTICLES, SENSIBLE BY NAKED EYES.

IN EARTH, NUMEROUS ATOMS, ELECTRONS, NUCLEONS, PARTICLE COMPOUNDS ARE CONSTRUCTED FROM

BIOFRIENDLY DIFFERENT TYPE PARTICLES, SUCH AS THERMAL PARTICLES, OR BIOFRIENDLY ELECTRIC PARTICLES, SONIC PARTICLES, LIGHT PARTICLES, ETC. ALSO SOME OTHER ELECTRONS NUCLEONS PARTICLE COMPOUND CONSTRUCTED FROM BIOHOSTILE PARTICLES, SUCH AS X-PARTICLE COMPOUNDS, GAMMA PARTICLE COMPOUNDS, HARSH THERMAL OR ELECTRICAL PARTICLE COMPOUND CONSTRUCTION ELECTRONS, NUCLEONS, AND ATOMS.

LIGHT FUNDAMENTAL PARTICLE CONSTRUCTED ATOMS, ELECTRONS, NUCLEONS, NANO-UNITS

LIGHT PARTICLE NANO-UNITS, LIGHT ENERGY PROVIDER ATOMS

CRYSTALINE ATOMS ARE ONE EXAMPLE FROM LIGHT-FUNDAMENTAL PARTICLE CONSTRUCTED ATOMS, CRYSTALINE ATOMS INTERNAL ELECTRONS -NUCLEONS SUBSYSTEM – UNITS PARTICLE- COMPOUNDS MOSTLY ARE BUILT FROM SUN ORIGIN LIGHT FUNDAMENTAL PARTICLE-COMPOUNDS, CRYSTALINE ELECTRONS-NUCLEONS MOSTLY CONSTRUCTED FROM RED, YELLOW, GREEN, BLUE, VIOLET, ETC. COLOR LIGHT PARTICLES – COMPOUNDS AS WELL AS MANY OTHER NON-DISCOVERED NON-SENSIBLE SUN-ORIGIN PARTICLE- COMPOUNDS WHICH THOSE ALSO ARE THE MAIN FUNDAMENTAL PARTICLES WHICH PARTICIPATED AT CONSTRUCTION OF THE CRYSTALNE ATOMS PARTICLE STRUCTURES.

THE LIGHT FUNDAMENTAL- PARTICLES ARE NEEDED FOR BUILDING THE PLANTS ELECTRONS – NUCLEONS LIGHT PARTICLE COMPOUND CONSTRUCTIONS, IN ANIMAL

SPECIES LIGHT FUNDAMENTAL PARTICLE PRODUCE INFORMATION IMAGE PARTICLE CLOUDS, WHICH THE LIGHT INFORMATION IMAGE PARTICLE –CLOUDS ARE MAIN ELEMENTS OF THOUGHT CURRENTS AS WELL AS THE LIGHT PARTICLE CLOUDS ALSO ARE USED IN CONSTRUCTIONS OF THE CNS ELECTRONS – NUCLEONS PARTICLE COMPOUND CONSTRUCTIONS, ALSO LIGHT FUNDAMENTAL PARTICLES INFORMATION IMAGE PARTICLE CLOUDS CURRENTS CONSTRUCT PSYCHE AND THOUGHT CURRENTS.

ELECTRIC FUNDAMENTAL PARTICLE CONSTRUCTED ELECTRONS, NUCLEONS, ATOMS, NANO-UNITS

ELECTRIC ENERGY PROVIDER ATOMS, THE NANO BATTERIES

THE ATOMS CONSTRUCTED FROM ELECTRIC FUNDAMENTAL PARTICLES ARE THE ELECTRIC PARTILCE ATOMS, THESE ATOMS ELECTRONS NUCLEONS PARTICLE COMPOUND CONSTRUCTIONS ARE MOSTLY BUILT FROM DIFFERENT CLASSES OF ELECTRIC FUNDAMENTAL PARTICLE, AND THE ELECTRIC FUNDAMENTAL PARTICLES OF THESE ELECTRIC MOLECULAR COMPOUND CONSTRUCTIONS INHERITANTLY POSSESSES QUANTUM ELECTRIC ENERGY CONTENS AT THE PARTICLES, THESE ARE QUANTUM QUANTITY ELECTRIC ENERGY SOURCES INSIDE ELECTRONS, NUCLEONS, NANO-UNIT, AND CAN GENERATE ELECTRIC ENERGY FOR ANY GIVEN ELECTRONS, NUCLEONS, NANO – SITES.

THROUGH DEGENERATIVE A. S. I. F.P. Mol. C. I. C. THESE

ELECTRIC FUNDAMENTAL-PARTICLES RELEASE FROM PARTICLE-COMPOUND CONSTRUCTIONS FORMS, INTO FREE ELECTRIC FUNDAMENTAL-PARTICLE FORMS, RELEASED ELECTRIC FUNDAMENTAL ARTICLES TRAVEL IN PARTICLE CURRENT CIRCULATION SYSTEMS, AND CAN BE USED AT INTER-ELECTRON INTER-NUCLEON CHEMICAL LAB.S AS ENERGY SOURCES OR CAN BE USED AS QUANTUM-ENERGY SOURCES AND ELECTRIC ENERGY PROVIDER IN INTERNAL-ATOM ELECTRIC BATTERIES OR USED IN CONSTRUCTION OF INTERNAL-ATOM INDUSTRIAL NANO-BATTERIES AND IN CONSTRUCTION OF ANY OTHER NANO- STRUCTURES.

ELECTRIC FUNDAMENTAL PARTICLES HAS UTMOST IMPORTANCE FOR ANIMAL ORGAN FUNCTIONS, ALSO ELECTRIC PARTICLE PROVIDER ATOMS WHICH THEIR PARTICLE CONSTRUCTIONS HAVE BEEN BUILT BY DIFFERENT ELECTRIC PARTICLE COMPOUNDS, ARE ALL CRITICALLY NEEDED FOR MAINTAINING AUTONOMOUS FUNCTIONS OF THE DIFFERENT BODY ORGAN AND SYSTEM, THEY PROVIDE ELECTRIC ENERGY FOR DIFFERENT QUANTUM –LOCATIONS AND CONDUCT PHYSICAL, CHEMICAL, BIOLOGICAL NANO- FUNCTIONS, AT DIFFERENT QUANTUM LOCATIONS.

THE PLANTS, ANIMAL SPECIES WITHOUT BIOFRIENDLY ELECTRIC PARTICLES AND CURRENTS, WHICH ARE PRODUCERS OF NANO-NEURAL CURRENTS AS WELL OERATERS OF NANO-ELECTRICAL TASKS FOR DIFFERENT BODY ORGANS AND SYSTEMS IN QUATUM LOCATIONS, CAN NOT EXIST, AND CAN NOT SURVIVE EVEN FOR A SECOND, WITHOUT ELECTRIC PARTICLES EXISTENCES.

IN INDUSTRIAL WORLD, THE ELECTRONS, NUCLEONS, ATOMS, AND ELECTRIC FUNDAMENTAL PARTICLES, WHICH THESE NANO-UNITS ELECTRIC PARTICLE COMPOUNDS HAVE BEEN CONSTRUCTED FROM HIGH ENERGY POWERFUL, HIGH WEIGHT, HIGH GRAVITON ELECTRIC PARTICLE CONSTRUCTIONS, THESE HIGH ELECTRIC ENERGY PARTICLES AND ELECTRIC PARTICLE COMPOUNDS HAVE BEEN USED TO PERFORM DIFFERENT KIND HIGH ELECTRIC ENERGY INDUSTRIAL ELECTRO-MECHANIC, THERMO-ELECTRIC, COMMERCIAL, INDUSTRIAL TASKS, AS WELL IN ELECTRIC GENERATOR PLANTS.

BY NO WAY THE ELECTRIC PARTICLES ENERGY AT EARTH ARE COMPARABLE TO OTHER VICIOUS HIGH ELECTRIC-ENERGY CONTENTS FUNDAMENTAL PARTICLES, AT HARSH PLANETS IN DISTANT GALAXIES NEXT TO BLACK HOLES THE ELECTRICAL ENERGY CONTNTS OF ELECTRIC FUNDAMENTAL PARTICLES ARE MILLIONS FOLDS STRONGER THAN EARTHS BIOFREINDLY WEAK ELECTRIC FUNDAMENTAL PARTICLES ELECTRIC ENERGY.

THERMAL FUNDAMENTAL PARTICLE CONSTRUCTED ELECTRONS, NUCLEONS, ATOMS

THERMAL PARTICLE CONSTRUCTED NANO-UNITS AND THERMAL ENERGY PROVIDERS

THE BIOFREINDLY WEAK THERMAL PARTICLES SUCH AS, INFRA - RED THERMAL FUNDAMENTAL PARTICLES, OR OTHER HARSH BIOHOSTILE THERMAL PARTICLES INTER INTO

INTERNAL ELECTRONS NUCLEONS SUBSYSTEMS UNITS CHEMICAL LAB.S, AND COMBINE WITH INTERNAL ELECTRON NUCLEON PARTICLE COMPOUNDS, PRODUCE INFRA-RED PARTILE COMPOUNDS, OR ANY OTHER THERMAL PARTICLE COMPOUNDS INSIDE ELECTRONS AND NUCLEONS THROUGH A. S. I. F.P. Mol. C.I.C.

THE INFRA-RED PARTICLES COMPOUNDS INSIDE ELECTRONS NUCLEONS ARE STORAGE OF HEAT PROVIDER PARTICLES INSIDE ELECTRONS NUCLEONS IN THERMAL PARTICLE COMPOUND FORMS, THESE THERMAL PARTICLE COMPOUNDS ARE THERMAL ENERGY PROVIDERS AND LIGHT ENERGY PROVIDER PARTICLES, READY TO DELIVER HEAT INTO ANY GIVEN QUANTUM LOCATIONS TO ACHIEVE NEEDED NANO-TASKS.

THE THERMAL PARTICLE COMPOUND CONSTRUCTED NANO-UNITS, ATOMS, ELECTRONS, NUCLEONS AT BIOLOGICAL WORLD ARE USED AS PROVIDERS AND GENERATORS OF THE QUATUM THERMAL-ENERGY SOURCES FOR PERFORMING INNORMOUS DIFFERENT KINDS BIOLOGICAL, CHEMICAL AND PHYSIOLOGICAL FUNCTIONS, UNDER DEGENERATIVE A. S. I. F.P. Mol. C.I.C. THE THERMAL PARTICLES RELEASE AS FREE HEAT PROVIDING FUNDAMENTAL PARTICLES, AND CAN PROVIDE THERMAL-ENERGY FOR ANY NANO-- LOCATION AT ANY GIVEN INTER ELECTRON NUCLEON QUANTUM LOCATIONS, TO DELIVER ON SITE HEAT –ENERGY FOR CHEMICAL INTERACTIONS, OR OTHER QUANTUM THERMO-DYNAMIC BIOLOGICAL – FUNCTIONS.

THE X- FUNDAMENTAL PARTICLE CONSTRUCTED ATOMS, GAMMA FUNDAMENTAL PARTICLE CONSTRUCTED ATOMS, AND OTHER HARSH BIOCIDAL CONSTRUCTED ATOMS ARE NOT DESCRIBED.

ATOM'S TURN OVER PHENOMENON, IS EQUAL TO EVOLUTION PATHS OF ATOM GENESIS

DEFINITION OF TURN OVER PHENOMENON:

PERIODICALLY DURING EACH ONE HALF LIFE OF ATOM, ELECTRONS, NUCLEON, THE ENTIRE PRE-EXISTING PARTICLE COMPOUNDS CONSTRUCTIONS OF DIFFERENT ELECTRONS, NUCLEONS, AND ATOM ALL REMOVED, AND IN THEIR PLACE ANOTHER EXACT COPY NEW EXACTLY THE SAME PARTICLE COMPOUNDS RECONSTRUCTED AS BEFORE, THIS IS PHENOMENONS OF THE ATONS, ELECTRONS, NUCLEONS TURN OVER PHENOMENO.

THE DESTRUCTIONS OF THE ELECTRONS, NUCLEONS, AND ATOMS PARTICLE COMPOUNDS ARE DONE THROUGH DEGENERATIVE A. S. I. P.F. Mol. C.I.C., AND THE REGENESIS OF ELECTRONS, NULEONS AND ATOMS PARTICLE COMPOUND CONSTRUCTIONS, TAKE PLACE THROUGH REGENERATIVE AUTONOMOUS SEQUENTIAL INTER FUNDAMENTAL PARTICLE MOLECULAR CHEMICAL INTERACTIONS CHAINS AND CYCLES.

THIS IS PERIODICAL CYCLES OF REGENERATIONS OF ELECTRONS NUCLEONS AND ATOMS PARTICLE COMPOUNDS, WHICH IT ALTERNATE DURING EACH GIVEN ONE HALF LIFE OF ELECTRONS, NUCLEONS AND ATOM WITH DEGENERATIVE CYCLES, DEGENERATIONS THAT TAKE PLACE IN EVERY ONE HAF LIFE OF ATOMS, THERE AFTER THIS PHENOMENON CONTINUES UNDER DEGENERATIVE A. S. I. F.P. Mol. C.I.C. AUTONOMOUSLY IN CYCLES, THROUGH THESE CHEMICAL INTERACTIONS ENTIRE ELECTRONS NUCLEONS AND ATOMS PARTICLE COMPOUDS STRUCTURES REMOVED, AND IN THEIR PLACE AOTHER NEW STRUCTURES COPIES OF THE PRE-EXISTING ORIGINALS RE-CONSTRUCTED.

THE REGENERATION CYCLES OF TURN OVER PHENOMENON OF ELECTRONS, NUCLEONS AND ATOMS ARE EQUAL TO EVOLUTION ERA RE-GENESIS OF THE ELECTRONS, NUCLEONS AND ATOMS DURING PAST MOLECULAR EVOLUTION ERA, THAT IT TOOK PLACE IN THE MATTERS OF MILLIONS YEARS UNTIL THE PRESENTLY EXISTING ELECTRONS, NUCLEONS, AND ATOMS CONSTRUCTED AS STABLE STRUCTURES WE HAVE TODAY.

THIS TURN-OVER PHENOMENON OCCUR SIMILARLY IN ANOTHER FIELDS AS WELL, AT MICRO-UNITS CONSTRUCTIONS, ALSO IN MACRO-UNITS FIELDS OF CONSTRUCTIONS AS WELL, FOR EXAMPLE THE WHOLE ANIMAL BODY AND CELL POPULATION PERIODICALLY WITHIN SPECIFIC GIVEN HALF LIVES OF TURN OVER PROCESS, THE ENTIRE OLD STRUCTURES REMOVED, AND IT ALTERNATE AND REPLACED WITH RE-CONSTRUCTIONS AND REGENESIS OF CELLS AND BODY

STRUCTURES WITH EXACT SAME COPIES AT REGENERATION CYCLES PERIDICALLY WITHIN SPECIFIC TIME INTERVALS.

MOLECULAR EVOLUTION

CYCLES OF AUTONOMOUS SEQUENTIAL CHEMICAL INTERACTIONS, AND CREATION OF LIFE

HOW DIFFERENT AUTONOMOUS SEQUENTIAL CHEMICAL INTERACTIONS CONSTRUCTED ATOMS, BIOMOLECULES, CELLS, LIVING THINGS:

PHASE –ONE OF EVOLUTION

1 – A. S. I. F.P. Mol. C.I.C. AND GENESIS OF NANO-UNITS:

THE ELECTRON-GENESIS, NUCELON-GENESIS, ATOM-GENESIS, ETC. (THE NANO-UNIT-GENESIS):

THE FUNDAMENTAL PARTICES ARE THE UNIT OF MATTER, AFTER INCEPTION OF PLANET EARTH, AT FIRST STEP THE AUTONOMOUS SEQUENTIAL INTER FUNDAMENTAL PARTICLE MOLECULAR CHEMICAL INTERACTION CYCLES AND CHAINS (A. S. I. F.P. Mol. C.I.C.) TOOK PLACE BETWEEN THE EARTH'S INDIGENOUS DIFFERENT FUNDAMENTAL PARTICLES, AND CONSTRUCTED DIFFERENT ELECTRONS, NUCLEONS. POSITRONS, QUARKS, NANO-UNITS AND ATOMS, ETC. THERE WAS NOT NANO UNITS CONSTRUCTIONS BEFORE THE FUNDAMENTAL PARTICLES AT EARTH.

THE PRESENTLY FEW DISCOVERED NANO-UNIT CLASSES SUCH AS ELECTRONS, NUCLEONS, ATOMS, POSITRONS, ETC. AT EARTH, ARE ONLY A FEW SMALL NUMBERS FROM NANO-UNITS CLASSES WHICH HAS BEEN DISCOVERED AT EARTH PRESENTLY, THERE ARE LARGE NUMBERS OF ANOTHER NANO-UNIT CLASSES, WHICH NOT BEEN DISCOVERED, AND NEEDS DISCOVERY IN FUTURE BY OTHERS.

PHASE –TWO OF EVOLUTION

2- A. S. I. N.-U. Mol. C.I.C. AND GENESIS OF BIOMOLECULES:

AT SECOND STEP THE AUTONOMOUS SEQUENTIAL INTER NANO-UNIT BASE MOLECULAR CHEMICAL INTERACTIONS CYCLES AND CHAINS (A. S. I. N.-U. Mol. C.I.C.) TOOK PLACE BETWEEN THE NEWLY CREATED DIFFERENT ATOMS AND ANOTHER NANO-UNITS, MOLECULAR STRUCTURES, AND DIFFERENT SIZES OF LARGER ATOM CONSTRUCTIONS, BIO-MOLECULES, ETC. CREATED DIFFERENT SIZES MOLECULAR STRUCTURES EXISTED IN GENERAL CHEMISTRY, ORGANIC CHEMISTRY AND BIO-CHEMISTRY TEXTS AND MANY MORE.

PHASE –THREE OF EVOLUTION

3 – A. S. I. C.B. Mol. C.I.C. AND GENESIS DIFFERENT CELLS, TISSUES, CREATURES, ETC:

THERE AFTER THE GENESIS OF DIFFERENT BIOMOLECULES, AND CELL, THE AUTONOMOUS SEQUENTIAL INTERNAL CELL

AND BIOMOLECULAR COMBINATIONS AND CHEMICAL INTERACTIONS CYCLES AND CHAINS, TOOK PLACE BETWEEN THE PRODUCED DIFFERENT BIOMOLECULES, CELLS, TISSUES, ETC. AND CONSTRUCTED MORE DIFFERENT OTHER SPECIES CELLS, TISSUES, ORGANS, SYSTEMS, ETC. AND PRODUCED INNORMOUS DIFFERENT SPECIES OF LIVING THINGS FROM PLANTS TO ANIMALS, WHICH THE LAST ONE OF THOSE MOST ADVANCED ANIMALS BELONG TO THE HUMAN SPECIES, THIS SPECIES TODAY OVER GROWN ALL OVER THE EARTH, FOR THEIR DESCRIPTIONS SEE BIOLOGICAL TEXTS BOOKS.

INTRODUCTION TO FUNDAMENTAL PARTICLE GENERAL CHEMICATRY COMBINING CRYSTALLINE INTERNAL ATOM LIGHT PARTICLE COMPOUNDS WITH EXOGENOUS EXTERNAL ATOM FUNDAMENTAL PARTICLES:

CRYSTALLINE ATOMS ARE LIGHT PARTICLE- COMPOUND CONSTRUCTED ATOMS, AND DIFFERENT CRYSTALLINE ELECTRONS –NUCLEONS HAS BEEN CONSTRUCTED FROM DIFFERENT KINDS NUMEROUS LIGHT FUNDAMENTAL PARTICLE COMPOUNDS CONSTRUCTIONS, UNDER REGENERATIVE A. S. I. F.P. Mol. C.I.C. DIFFERENT KINDS EXOGENOUS INCIDENT INCOMING FUNDAMENTAL PARTICLES SUCH AS X- FUNDAMENTAL PARTICLES, ELECTRIC- FUNDAMENTAL PARTICLES, GAMMA-PARTICLES, ETC. ONE BY ONE COMBINED ONE TEST AFTER OTHER, AND EACH GIVEN CHEMICAL COMBINATIONS TEST PRODUCED DIFFERENT KIND PARTICLE BY PRODUCTS (P. B. P.) SHOWED DIFFERENT A. S. I. F.P. Mol. C.I.C. BETWEEN DIFFERENT FUNDAMENTAL PARTICLE MOLECULAR STRUCTURES CAUSE PRODUCTION OF DIFFERENT KIND PARTICLE –BY- PRODUCTS WHICH IN ALL TESTS THE RESULTS ARE A. S. I. F.P. Mol. C.I.C. –SPECIFIC, AND IN DIFFERENT TESTS THE RESULTS ARE DIFFERENT FROM EACH OTHERS.

UNDER FOLLOWING TESTS DIFFERENT INDOGENOUS INTERNAL ELECTRON-NUCLEON LIGHT -PARTICLE COMPOUNDS OF A GIVEN CRYSTALINE COMBINED WITH DIFFERENT EXTERNAL ATOM EXOGENOUS FUNDAMENTAL PARTICLES ONE TEST AFTER OTHER UNDER SEPARATE TESTS AS FOLLOWING:

1-   IN FIRST TEST COMBINATION OF GIVEN EXOGENOUS INCIDENT ELECTRIC - PARTICLES WITH GIVEN COMPATIBLE GROUPS OF INTER ELECTRON-NUCLEON LIGHT PARTICLE COMPOUNDS OF GIVEN CRYSTALINE ACHIEVED THROUGH A. S. I. F.P. Mol. C.I.C. UNDER SPECIFIC TEST CONDITIONS AND PRODUCED SPECIFIC GIVEN PARTICLE-BY-PRODUCTS (P.B.P.) SPECIFIC OF MOLECULAR COMBINATIONS.

2 – IN SECOND TEST GIVEN EXOGENOUS INCIDENT X- FUNDAMENTAL PARTICLES COMBINED WITH OTHER GROUPS OF COMPATIBLE GIVEN INTERNAL ELECTRONS –NUCLEONS LIGHT PARTICLE COMPOUNDS OF CRYSTALINE UNDER SPECIFIC TEST CONDITIONS AND PRODUCED SPECIFIC GIVEN PARTICLE –BY – PRODUCTS (P.B.P.) SPECIFIC FOR THIS A. S. I. F.P. Mol. C.I.C.

3 – IN THIRD TEST GIVEN SPECIFIC INCIDENT EXOGENOUS GAMMA FUNDAMENTAL PARTICLES COMBINED WITH DIFFERENT COMPATIBLE INTERNAL ELECTRON –NUCLEON LIGHT – PARTICLE COMPOUNDS OF CRYSTALINE UNDER SPECIFIC TEST CONDITIONS AND PRODUCED TEST SPECIFIC PARTICLE BY PRODUCTS UNDER GIVEN TEST CONDITIONS THROUGH A. S. I. F.P. Mol. C.I.C.

THE ABOVE THREE TESTS, THREE DIFFERENT EXOGENOUS PARTICLES CHEMICALLY COMBINED WITH THREE DIFFERENT KINDS INTERNAL ELECTRON NUCLEONS LIGHT PARTICLE COMPOUNDS, AND PRODUCED THREE DIFFERENT COLOR VISIBLE LIGHT PARTICLES, LIGHT PARTICLES VISIBLY SCAPED AIRBORN.

IT IS OBVIOUS THE CHEMICAL COMBINATION OF GAMMA PARTICLES WITH GIVEN LIGHT –PARTICLE-COMPOUNDS PRODUCE GAMMA-PARTICLE COMPOUND, PLUS A GIVEN SPECIFIC KIND - COLOR LIGHT PARTICLE WHICH THIS RELEASED LIGHT PARTICLE IS SPECIFIC FOR THIS GIVEN TEST AND COMBINING LIGHT PARTILE COMPOUND STRUCTURE, THE RELEASED LIGHT PARTICLE SCAPE FREE AIRBORN AND SENSED WHEN A. S. I. F. P. Mol. C.I.C. ACCOMPLISHED, THE P. B. P. PRODUCED UNDER THIS CHEMICAL COMBINATION IS PRODUCTION OF SPECIFIC COLOR AND KIND LIGHT PARTICLE PLUS GAMMA- PARTICLE COMPOUNDS WHICH ARE USED IN NEW CONSTRUCTIONS OF ELECTRONS- NUCLEONS (PARTICLE –COMPOUND NEO-GENESIS) SPECIFIC OF THIS TEST.

THE A. S. I. F.P. Mol. C.I.C. RESULTS OF ELECTRIC –FUNDAMENTAL PARTICLES COMBINATIONS WITH LIGHT-PARTICLE-COMPOUNDS OF CRYSTALINE PRODUCE PRODUCTION OF ELECTRIC- PARTICLE –COMPOUND PLUS ANOTHER KIND AND COLOR TEST SPECIFIC LIGHT PARTICLE WHICH AGAIN ARE SENSABLE AND SCAPE VISIBLE TO AIR.

THE CHEMICAL COMBINATIONS RESULTS WITH X-PARTICLES

PRODUCE DIFFERENT KIND TEST SPECIFIC RESULTS, SOME TESTS MAY RELEASE BRIGHT WHITE LIGHT PARTICLES, THE OTHER RED OR BLUE COLOR LIGHT PARTICLES, ETC.

ABOVE TESTS CAN BE REPEATED WITH REMARKABLY LARGE NUMBERS OF ANOTHER KIND ATOMS OTHER THAT CRYSTALINE OR DIFFERENT KIND OTHER CRYSTALINE ATOMS AND THE ALL TEST RESULTS ALL ARE INTERACTION –SPECIFIC WITH NO CHANGE, EVEN SOME TIMES THE LIGHT PARTICLES WHICH ARE NOT EXIST IN EARTH MAY SCAPE VISIBLE, IN THE SAME TIMES MANY OF THE OTHER LIGHT PARTICLES OR NON-LIGHT PARTICLES OF ANOTHER KINDS OBVIOUSLY ARE NOT DETECTABLE PRESENTLY.

IN THESE TESTS THERE ARE LARGE NUMBERS OF NON-DISCOVERED FUNDAMENTAL PARTICLE ALSO PARTICIPATE AT INTERACTIONS AS WELL AS RELEASED BY THESE TESTS WHICH WE HAVE NO ABILITY TO EVALUATE THOSE AT PRESENT TIME.

ELECTRON -CYCLES
COMBINING AN INTERNAL ELECTRON PARTICLE COMPOUND WITH INCIDENT EXTERNAL ATOM PARTICLE MOLECULE: OR PHENOMENON OF TRIGGERING CENTRIFUGAL AND CENTRIPETAL ELECTRON'S TRANSITION CYCLES

SHINE ONE MOLECULAR –UNIT FUNDAMENTAL PARTICLE (PHOTON) INTO A GIVEN ELECTRONS CHEMICAL LAB.S, AND

COMBINE EXOGENOUS PHOTON (Y. –F.P.) WITH INTERNAL ELECTRON PARTICLE- COMPOUNDS (F.P. _ COMP.) IN GIVEN ORBIT, COMBINE ONE MOLECULAR UNIT INTER-ELECTRON INDOGENOUS PARTICLE-COMPOUND WITH ONE MOLECULAR-UNIT EXOGENOUS FUNDAMENTAL PARTICLE (Y. –F.P) UNDER A. S. I. F. P. Mol. C.I.C., WILL PRODUCE PHOTON – COMPOUND (Y. F.P. _ COMP.) THIS COMBINATION WILL CAUSE THE WEIGHT OF PARTICLE COMPOUND INCREASE EQUAL TO THE COMBINING PHOTON WEIGHT, EQUATION IS AS FOLLOWING.

$$Y.\text{-}F.P. + F.P.\_COMP. \xrightarrow{} Y.-F.P.\_COMP.$$

$$\xleftarrow{}$$

INCREASE IN Y. F.P. _ COMP. WEIGHT, IS EQUAL TO INCREASE IN THE ELECTRON'S WEIGHTS, THESE INCREASES OF ELECTRONS OR PHOTON -COMPOUND WEIGHTS, ARE EQUALS TO THE WEIGHT AMOUNTS OF COMBINING PHOTON. THIS A. S. I. F.P. Mol. C.I.C. WILL CAUSE THE ELECTRONS WEIGHTS INCREASE EQUAL TO WEIGHT OF COMBINING PHOTON. INCREASE IN ELECTRON'S WEIGHT IS PROPORTION TO INCREASE IN ELECTRON'S ENERGY AMOUNT, AND ELECTRON'S GRAVITON FORCE INCREASE.

THIS INCREASE IN ELECTRON'S WEIGHT, GRAVITON, ENERGY, WILL CAUSE THE ELECTRON TO TRANSIT FROM LOW ENERGY ELECTRON ORBIT, INTO HIGH ENERGY ELECTRON ORBIT LEVEL, THIS PHENOMENON IS CENTRIFUGAL ELECTRON TRANSIT CYCLE, OR IT IS ELECTRON'S CENTRIFUGAL

TRANSITION CYCLE IN DIRECTION OF INDEPENDENT AND ELECTRON TRANSIT TOWARD PERIPHERAL ORBITAL DIRECTIONS.

MOST OF FUNDAMENTAL PARTICLE CHEMICAL COMBINATIONS ARE REVERSIBLE, THAT MEANS ORIGINAL CHEMICAL INTERACTIONS REVERSE ITS ORIGINAL EQUATIONS DIRECTION SPONTANEOUSLY BACKS INTO OPPOSITE DIRECTION OF ORIGINAL REGENERATIVE A.S. I. F.P. Mol. C.I.C.

REGENERATIVE A. S. I. F.P. Mol. C.I.C. CAUSED CONSTRUCTION OF PHOTON PARTICLE COMOUND INCREASE IN ELECTRON WEIGHT, THE DEGENERATIVE A. S. I. F.P. Mol. C.I.C. CAUSE BREAK DOWN OF PHOTON COMPOUND INTO CONSTRUCTING SMALLER MOLECULES, AND RELEASE OF PHOTON BACK AIRBORN, THIS PROCESS CAUSE DECREASE IN WEIGHT OF ELECTRON AND PARTICLE COMOUNDS. IN CONSEQUENCE THE ELECTRONS HAS TO TRANSIT BACK INTO LOW ENERGY LEVEL ORBITS AT REVERSE DIRECTION, THIS PHENOMENON IS CENTRIPETAL ELECTRON- TRANSIT TOWARD DEPENDENCE, OR ITS FINAL TRAVEL INTO COMBINATION WITH NUCLEONS UNDER ATTRACTION GRAVITON FORCES, THIS IS ELECTRON'S CERTRIPETAL TRANSIT CYCLES, THE ELECTRON TRANSIT TOWARD CENTRAL DIRECTIONS.

REGENERATIVE A. S. I. F.P. Mol. C.I.C. PRODUCE GENESIS OF PHOTON _ COMPOUNDS (Y. – F. P. _ COMP.) INSIDE ELECTRONS SUBSYSTEM- UNITS, WHICH CONSTRUCT INTERNAL ELECTRON PHOTON COMPOUND CONSTRUCTIONS OF ELECTRONS, THIS PHENOMENON IS NEO- GENESIS OF

FUNDAMENTAL PARTICLE COMPOUNDS, CAUSING NEO-GENESIS OF ELECTRONS.

INTRODUCTION TO ORGANIC CHEMISTRY AND BIOCHEMISTRY SCIENCES OF FUNDAMENTAL PARTICLES

PHENOMENON OF INTELLIGENT ELECTRON GENESIS, PHE-NOMENON OF INTELLIGENT NUCLEON GENESIS

GENESIS OF INTELLIGENT ELECTRONS, NUCLEONS, PRO-DUCED, THROUGH USE OF S. Y. − F.P. − I.I. − P. cl. TO CONSTRUCT CNS, ELECTRONS − NUCLEONS, PARTICLE COMPOUND CONSTRUCTIONS, THROUGH THE BUILDING ELECTRONS NULEONS PARTICLE COMPOUND CONSTRUCTIONS BY USE OF SONIC −LIGHT PARTICLE CLOUDS AND CONSTRUCTIONS OF INTELLIGENT S. Y. −F.P. − I.I.- P.cl. _COMPOUNDS, INSIDE CNS ELECTRONS AND NUCLEONS, WHICH IT IS ORIGIN OF THOUGHT CURRENT SYSTEMSS.

UNDER REGENERATIVE A. S. I. F.P. Mol. C.I.C. COMBINE EXOGENOUS S. Y.- F.P. − I.I. − P. cl. WITH CNS INTER ELECTRON- NUCLEON PARTICLEE- COMPOUNDS AND PRODUCE INTERNAL ELECTRON-NUCLEON F.P. − I.I. − P.cl. _ COMPOUNDS.

ENVIRONMENTAL EXISTING THINGS PRODUCED S. Y. - F.P. − I.I. − P. cl. AIRBORN TRANSIT INTO RECIPIENTS CNS INTER CNS ELECTRONS −NUCLEONS, COMBINE WITH INTER ELECTRON-NUCLEON PARTICLE COMPOUNDS INSIDE SUBSYSTEM −UNITS CHEMICAL LAB.S, UNDER A. S. I. F.P. Mol. C.I.C.,

AND PRODUCE S. Y. - F.P. – I.I. – P.cl. __COMPOUNDS, THESE VIRTUAL PARTICLE CLOUD COMPOUNDS CONSTRUCT CNS ELECTRONS – NUCLEONS CONSTRUCTIONS,

THIS IS STORAGE AND ACCUMULATIONS OF ENVIRONMENT SUBJECTS VIRTUAL PARTICLE CLOUDS, ABOUT ALL KIND SOCIAL POLITICAL FINANCIAL EDUCATIONAL OR SCIENCE MATTERS WHICH STORES INSIDE BRAINS ELECTRONS AND NUCLEONS IN PARTICLE –COMPOUN FORMS, LIVING THINGS PILE UP OF DIFFERENT KINDS VIRTUAL PARTICLE CLOUDS COMPOUNDS, INSIDE THEIR BRAINS, AND CONSTRUCT THEIR CNS ELECTRONS –NUCLEONS WITH S.Y. – F.P. –I.I. – P.cl. THROUGH BUILDING ELECTRONS NUCLEON CONSTRUCTIONS WITH USE OF VIRTUAL PARTICLE CLOUDS COMPOUNDS, AND THIS IS PILING UP OF DIFFERENT KINDS KNOWLEDGE- INTELLIGENCE –SCIENCE ETC.IN BRAIN LIFE LONG, IT IS PHENOMENON OF LEARNING GAINING KNOWLEDGE ABOUT DIFFERENT MATTERS, ABOVE PHENOMENON IS STORAGE OF S.Y. – F.P. – I.I.- P.cl. INSIDE THE CNS ELECTRONS NUCLEONS IN PARTICLE CLOUD COMPOUND FORMS.

COMBINING EXOGENOUS S. Y. - F.P. – I.I. – P. cl. WITH INTER-ELECTRON-NUCLEON F. P. _ COMPOUNDS OR S.Y. – F.P. – I.I.- P. cl. STORAGE INSIDE ELECTRONS NUCLEONS AND S. Y. –F.P. – I.I. – P. cl. RETRIEVAL FROM INTER – ATOM PARTICLE COMPOUND TO OUSIDE PARTICLE CLOUD STORAGE, AND PARTICLE CLOUD RETRIEVAL, FROM ELECTRONS NUCLEONS

THIS CHEMICAL COMBINATION TAKE PLACE BETWEEN GIVEN EXTERNAL ATOM EXOGENOUS SONIC-LIGHT FUNDAMENTAL PARTICLE INFORMATION- IMAGE PARTICLE CLOUDS (S. Y. – F.P. – I.I. – P. cl.) COMBINING WITH CNS INTERNAL ELECTRON –NUCLEON PARTICLE COMPOUNDS, AND COMBINATION PRODUCE S. Y. – F.P. – I.I.- P. cl. _ COMPOUND (OR VIRTUAL PARTICLE CLOUD COMPOUND) INSIDE CNS RECIPIENT ELECTRONS- NUCLEONS, OR INSIDE NON-LIVE ATOMS SUCH AS SILICON ELECTRONS-NUCLEONS WHEN COMBINING WITH NON-LIVE ELECTRONS-NUCLEONS.

THESE CHEMICAL COMBINATIONS PRODUCE INTELLECTUAL VIRTUAL PARTICLE CLOUD COMPOUNDS CONSTRUCTIONS, AND CONSTRUCT RECIPIENT'S CNS ELECTRONS –NUCLEONS PARTICLE COMPOUND CONSTRUCTIONS, THROUGH COMBINATIONS OF VIRTUAL – PARTICLE CLOUDS (INTELLECTUAL

S. Y. – F.P. – I.I. – P. cl. MOLECULAR STRUCTURES) WITH CNS INTERNAL ATOM PARTICLE COMPOUNDS,

THESE S. Y. – F.P. –I.I. –P. cl. CONTAIN AUDIO-VISUAL INFORMATIONS IMAGES KNOWLEDGE ABOUT OUTSIDE WORLDS ENVIRONMENTAL SUBJECTS, THESE INFORMATION IMAGE PARTICLE –CLOUDS ARE PARTICLE –CLOUD COPIES OF ORIGINAL SUBJECTS, THE VIRTUAL CLOUDS ARE EXACT COPIES OF ORIGINAL SUBJECTS PARTICLE COPIES, AND MADE IN PARTICLE – CLOUD –FORMS.

WHEN AIRBORN VIRTUAL PARTICLE CLOUDS INTER INTO BRAINS AND CONSTRUCT S. Y. –F.P. – I.I. – P. cl. _ COMPOUND INSIDE RECIPIENT CNS ELECTRON-NUCLEON, THIS PHENOMENON IS EQUAL TO STORING ORIGINAL SUBGECTS VIRTUAL PARTICLE COPIES, INSIDE CNS ELECTRONS NUCLEONS AND ATOMS, WHEN THESE VIRTUAL PARTICLE CLOUDS MOVES IT FEEL EXACTLY THE ORIGINAL SUBJECTS AND ORIGINAL MATTERS MOVED INSIDE THEIR CNS ELECTRONS NUCLEONS, THIS IS THE BIOLOGICAL FEELINGS THAT PRODUCE ANIMALS THOUGHT CURRENTS AND PSYCHE.

THE VIRTUAL –PARTICLE CLOUDS ACTIONS ALL ARE BIOLOGICALLY SENSED AS IDENTICAL TO ORIGINAL SUBJECTS ACTS AND FUNCTIONS, THIS IS PHENOMENON OF PSYCHE –GENESIS AND THOUGHT CURRENT –GENESIS PHENOMENON, WHICH VIRTUAL PARTICLE CLOUD IS IDENTICAAL TO THE REAL ORIGINAL SUBJECTS AND MATTERS, BIOLOGICALLY LIVING THINGS SENSE ALL OF THESE PHENOMENONS INSIDE THEIR BRAINS AND KALL IT THE SOUL.

THEREFORE S.Y. –F.P. – I.I. – P. cl. COMPOUNDS (VIRTUAL –CLOUDS COMPOUNDS) STORE DIFFERENT ENVIRONMENTAL AUDIO-VISUAL INFORMATIONS IMAGES, KNOWLEDGES ABOUT DIFFERENT OUT SIDE EXISTING THINGS AND MATTERS INSIDE CNS ELECTRONS – NUCLEONS.

THE CONSTRUCTIONS OF CNS ELECTRONS- NUCLEONS WITH S. Y. – F.P. – I.I. – P. cl. COMPOUNDS IN REALITY ARE SIMILAR TO A LIBERARY OR ENCYCLOPEDIA EVERY THING IS STORED THERE AT INSIDE YOU CAN FIND RELATIVELY.

THESE S. Y.- F.P. –I.I. – P.cl. ALSO CAN BE RELEASED FREE AND RETRIEVED TO OUT OF S. Y. – F.P. I.I. – P. cl. _ COMPOUNDS, THROUGH DEGENERATIVE A. S. I. F.P. Mol. C.I.C., WHICH THIS IS RETRIEVAL PHENOMENON OF PARTICLE CLOUDS TO OUSIDE WORLD.

S. Y. – F.P. – I.I. – P. cl. STORAGE PHENOMENON INSIDE
ELECTRONS-NUCLEONS
STORAGE OF PARTICLE CLOUDS INSIDE
ELECTRONS -NUCLEONS,

BRIEFLY EXOGENOUS EXTERNAL S. Y. – F.P. – I.I. – P. cl. (VIRTUAL CLOUDS) THROUGH SENSORY PARTICLE TRANSMISSION ROUTES INTER INTO DIFFERENT CNS SENSORY CENTERS ELECTRONS NUCLEONS, UNDER A.S. I. F.P. Mol. C.I.C. COMBINE WITH INTERNAL ELECTRON- NUCLEON PARTICLE COMPOUNDS, (UNDER REGENERATIVE AUTONOMOUS SEQUENTIAL IN PARTICLE – CLOUD CHEMICAL INTERACTION CYCLES), AND COMBINATION PRODUCE S. Y. - F.P.

– I.I. – P.cl. _ COMPOUND (VIRTUAL- CLOUD COMPOUND), AND STORE ENVIRONMENTAL EVENTS INFORMATIONS IMAGES IN S. Y. - F.P. – I.I. – P.cl. – COMPOUNDS FORMS INSIDE ELECTRONS- NUCLEONS OF RECIPIENT ATOMS, SUCH AS NON- LIVE SILICON ATOMS IN NON-LIVE WORLD OF ATOMS, OR IT STORE INTELLIGENT MOLECULAR S. Y. – F.P. – I.I. – P. cl. – COMPOUNDS (VIRTUAL CLOUD COMPOUNDS) THE INFORMATION AND ENVIRONMENTAL KNOWLEDGES INIDE CNS ELECTRONS NUCLEONS LIVING THINGS SPECIES.

THIS PHENOMENON IS RECORDING OR STORING OR PROGRAMMING OF OUTSIDE ENVIRONMENTAL INCIDENTS INFORMATIONS- IMAGES INSIDE ELECTRONS- NUCLEONS, OR SIMPLY RECORDING KNOWLEDGE INSIDE ELECTRONS- NUCLEONS LIVING THINGS SUCH AS ANIMALS OR INSIDE NON LIVE ATOMS SUCH AS SEMI-CONDUCTORS ATOMS.

ABOVE IS LEARNING PHENOMENON AND KNOWLEDGE – ACCUMULATION PROCESS, AND ARE USED IN HOME BY PARENTS, OR AT SCHOOLS TO TEACH HARD CORE SCIENCE OR ANY THING ELSE IN CLASS ROOMS FROM KINDER GARDEN UP TO THERE AFTER POSTGRADUATE TEACHING SYSTEMS OF DIFFERENT UNIVERSITIES, OR AT DIFFERENT TRAINING FILEDS AND EDUCATIONAL PROGRAMS STARTS FROM KINDER GARDENS AND CONTINUE FOR ALL LIFE, THESE USED IN SOCIAL POLITICAL RELIGIOUS OR ANY OTHER FILED ETC. FOR LEARNING AND ACCUMULATIONS OF ALL DIFFERENT LEGAL OR ILLEGAL TRAININGS.

THESE STORED KNOWLEDGES ALL HAVE ONE HALF LIFE

TIME MOLECULAR CONSTRUCTION STABILITIES, AFTER EACH ONE HALF LIFE TIME THEIR QUANTITIES DECLINE APPROXIMATELY TO ONE HALF OF ACCUMULATE QUANTITY, THESE STORAGE S.Y. –F.P.- I.I. – P. cl. EVEN AFTER DEMISE OF LIVING THINGS STAY INSIDE ELECTRONS –NUCLEONS ACCORDING THEIR HALF LIFE SCHEDULES AND GRADUALLY DISINTEGRATE UNDER REVERSE A. S. I. F.P. Mol. C.I.C.

RELEASE AND RETRIEVAL OF PARTICLE – CLOUDS
TO OUT OF PARTICLE -COMPOUNDS

PHENOMENON OF RETRIEVAL OF S. Y. – F.P. I.I. – P. cl. TO OUTSIDE CNS ELECTRONS-NUCLEONS

THESE STORED S. Y. - F.P. – I.I. – P.cl. – KNOWLEDGES IN PARTICLE –CLOUD MOLEULAR FORMS AT LATER TIMES CAN BE RELEASED TO OUT OF ELECTRONS- NUCLEONS UNDER DEGENERATIVE A. S. I. F.P. Mol. C.I.C. AS FREE RELEASED PARTICLE –CLOUDS AND THEIR INFORMATIONS- IMAGES OR THEIR MOLECULAR KNOWLEDGES CAN BE DEMONSTRATED AND VISUALIZED ON T.V. SCREENS AGAINS IN AUDIO-VISUAL FORMS.

NANO – UNIT CHEMISTRY
NANO-UNIT CHEMISTRY AND CHEMICAL
COMBINATIONS BETWEEN NANO-UNITS
ELECTRON-GENESIS, PROTON-GENESIS,
NEUTRON-GENESIS, AND ATOM – GENESIS
CHEMICAL COMBINATIONS BETWEEN ELECTRONS
AND NUCLEONS
NANO-UNITS, CHARACTORS OF NANO-UNITS.

DURING MOLECULAR EVOLUTION A. S. I. F.P. Mol. C.I.C. BETWEEN DIFFERENT FUNDAMENTAL PARTICLES AND PARTICLE COMPOUNDS CONSTRUCTED PRIMARY NANO-UNITS ORGANS AND SYSTEMS SUCH AS SUBSYSTEM – UNITS CHEMICAL LAB.S, PARTICLE TRANSPORTATIONS SYSTEMS, SUBSYSTEMS – UNITS INTELLIGENCE SYSTEM CENTERS, ETC., ALSO CONTINUATION OF DIFFERENT OTHER CHEMICAL COMBINATIONS BETWEEN PRODUCED NANO-UNITS PRODUCED MORE DIFFERENT QUANTUM SIZE ANOTHER CLASS STRUCTURES UNDER AUTONOMOUS SEQUENTIAL INTER NANO-UNIT CHEMICAL INTERACTION CYCLES, GRADUALLY DIFFERENT NANO-UNITS SUCH AS ELECTRONS, POSITRONS, QUARKS OR SYSTEMS AND SUBSYSTEM STRUCTURES, PROTONS, NEUTRONS, ATOMS, BARYONS, LEPTONS, MESONS, ETC. ONE AFTER OTHER PRODUCED, WHICH ALL

ARE DIFFERENT SIZES DIFFERENT CLASSES NANO- UNITS THEIR FUNDAMENTAL TOTAL POPULATIONS FROM ONE NANO-UNIT TO OTHER VARIES REMARKABLY.

TRANSFORMATION OF NANO-UNITS AND MUTATIONS OF NANO-UNITS FROM ONE TO OTHER THOSE ARE ALSO ANOTHER SUBJECTS WHICH OCCUR FREQUENTLY, HERE MOSTLY DIFFERENT AUTONOMOUS SEQUENTIAL INTER NANO-UNIT CHEMICAL INTERACTIONS CYCLES AND CHAINS (A. S. I. N.-U. C.I.C.) BETWEEN DIFFERENT NANO-UNITS BRIEFLY EXPLAINED AS FOLLOWING.

REGARDING PHYSIOLOGICAL FUNCTIONS IN NANO-UNITS AND FUNDAMENTAL PARTICLES, THE INTER-PARTICLE BEHAVIORS SUCH AS HOSTILITIES OF ONE PARTICLE TO OTHER, OR COOPERATION BETWEEN PARTICLES, THAT ONE PARTICLE DAMAGE ACTIVITIES OF OTHER, AND MAKE OTHER PARTICLE MALFUNCTION, OR PARTICLES HELP EACH OTHER AND COOPERATE, ETC. NOW PARTICLES ARE USED IN CYBER-ATTACKS OF ONE PARTICLE AGAINST OTHERS, ONE PARTICLE DISABLE THE OTHER, THESE ARE NEW WEAPONS.

PROTON-NEO-GENESIS,
PROTON-GENESIS
NANO-UNIT CHEMISTRY
THREE INTER- NANO-UNIT INTERACTIONS
AS FOLLOWING PRODUCE PROTONS:

1 - UNDER DEGENERATIVE AUTONOMOUS SEQUENTIAL INTER NANO-UNITS CHEMICAL INTERACTION CYCLES AND

CHAINS (A. S. I. N. – U. C.I.C.): FROM CHEMICAL CONSTRUCTION OF ONE NEUTRON, REMOVE ONE ELECTRON AND ONE ELECTRON ANTI-NEUTRINO, WILL PRODUCE ONE PROTON. THIS IS PROTON-GENESIS UNDER A. S. I. N.-U. C.I.C.:

NEUTRON - ELECTRON - Electron Antineutrino ----------
→ PROTON

←-------------

2 – UNDER REGENERATIVE A. S. I. N. – U. C.I.C.: COMBINE ONE POSITRON AND ONE ELECTRON NEUTRNO, WITH ONE NEUTRON, WILL PRODUCE ONE PROTON. THIS IS PROCESS OF PROTON –GENESIS, UNDER A. S. I. N. – U. C.I.C.:

NEUTRON + POSITRON + Electron Neutrino ------------
-→ PROTON

←----------------

3 - UNDER DEGENERATIVE AND REGENERATIVE AUTONOMOUS SEQUENTIAL INTER NANO-UNIT CHEMICAL INTERACTION CYCLES: FROM ONE NEUTRON CONSTRUCTION, REMOVE ONE ELECTRON, ALSO ADD ONE ELECTRON-NEUTRINO, WILL PRODUCE ONE PROTON.

NEUTRON - ELECTRON + Electron Neutrino --------
→ PROTON

THE DIFFERENT NEUTRONS, PROTONS, ELECTRONS, POSITRONS, NEUTRINO, ANTINEUTRINO INTERACTIONS WITH EACH OTHER THROUGH THE SEQUENTIAL AUTONOMOUS INTER NANO-UNIT CHEMICAL INTERACTIONS UNDER FUNDAMENTAL-PARTICLE EVOLUTION ORDERS AND PATHS CAUSE PRODUCTIONS OF ANOTHER NANO-UNITS, DIFFERENT CHEMICAL INTERACTIONS BETWEEN DIFFERENT NANO-UNITS CREATE NEW PROTONS, NEUTRONS, ATOMS AND ELECTRONS, ETC., THESE ARE PHENOMENONS OF NANO-UNIT NEO- GENESIS.

NEUTRON GENESIS
NEUTRON NEO – GENESIS
(NANO-UNITS CHEMISTRY)

THE FOLLOWING THREE NEUTRON –GENESIS INTERACTIONS, UNDER A. S. I. N.- U. C.I.C. PRODUCE NEUTRON:

1 - UNDER DEGENERATIVE AUTONOMOUS SEQUENTIAL INTER NANO –UNIT CHEMICAL INTERACTION CYCLES AND CHAINS: REMOVE ONE POSITRON, AND ONE NEUTRINO, FROM ONE PROTON CONSTRUCTION, WILL PRODUCE ONE NEUTRON. THIS IS NEUTRON GENESIS PROCESS.

PROTON - POSITRON - Electron Neutrino -----------
            → NEUTRON

          ←------------

2 - THROUGH REGENERATIVE INTER-NANO-UNIT AUTONOMOUS SEQUENTIAL CHEMICAL INTERACTIONS: COMBINE ONE PROTON WITH ONE ELECTRON, AT SAME TIME REMOVE ONE ELECTRON ANTI-NEUTRINO, WILL PRODUCE ONE NEUTRON. UNDER DEGENERATIVE INTER NANO-UNIT CHEMICAL INTERACTIONS. THIS IS NEUTRON GENESIS:

PROTON + ELECTRON - Electron Antineutrino -----
-------→ NEUTRON

←--------------

3 - UNDER REGENERATIVE AUTONOMOUS SEQUENTIAL INTER-NANO-UNIT CHEMICAL INTERACTIONS:

COMBINE ONE PROTON WITH ONE ELECTRON, AND REMOVE ONE ELECTRON NEUTRINO, WILL PRODUCE NEUTRON. THROUGH DEGENERATIVE A. S. I. N.- U. C.I.C. AND THIS IS PHENOMENON OF NEUTRON GENESIS.

PROTON + ELECTRON - Electron-Neutrino ------
---------→ NEUTRON

←-------------------

NANO-UNIT CHEMISTRY

ELECTRON-GENESIS, POSITRON-GENESIS

FOLLOWING ARE COUPLE EXAMPLES FROM PROCESS OF ELECTRON GENESIS, POSITRON GENESIS, UNDER A. S. I. N. – U. –C.I.C.:

1 - ELECTRON- GENESIS:

UNDER DEGENERATIVE – REGENERATIVE A. S. I. N. –U. C.I.C. FROM ONE NEUTRON CONSTRUCTION REMOVE ONE PROTON, AND ADD ONE ANTI-NEUTRINO, WILL PRODUCE ONE ELECTRON. THIS IS PHENOMENON OF ELECTRON-GENESIS.

NEUTRON - PROTON + Electron Anti-Neutrino ------------------→ ELECTRON

←----------------

2 – ELECTRON – GENESIS:

UNDER NUMEROUS DEGENERATIVE AND REGENERATIVE AUTONOMOUS SEQUENTIAL INTER-NANO-UNIT CHEMICAL INTERACTIONS: FROM ONE NEUTRON CONSTRUCTION REMOVE ONE PROTON, AND ADD ONE NEUTRINO, WILL PRODUCE ONE ELECTRON. THIS PHENOMENON IS ELECTRON-GENESIS,

NEUTRON - PROTON + Electron Neutrino -----------
→ ELECTRON

←------------

CHEMISTRY OF NANO -UNITS
POSITRON – GENESIS

UNDER DEGENERATIVE AUTONOMOUS SEQUENTIAL INTER NANO –UNIT CHEMICAL INTERACTIONS CYCLES: FROM ONE PROTON CONSTRUCTION REMOVE ONE NEUTRON, ALSO REMOVE ONE ELECTRON NEUTRINO, WILL PRODUCE ONE POSITRON. THIS IS PHENOMENON OF POSITRON - GENESIS.

PROTON - NEUTRON   -   Electron Neutrino   -------------
     ---→   POSITRON

    ←--------------------

FUNDAMENTAL PARTICLE'S BIOLOGICAL
SCIENCES, IN PLANTS
FUNDAMENTAL PARTICLE'S BIOLOGICAL
CHEMISTRY SCIENCES, IN PLANTS
BIOCHEMISTRY OF FUNDAMENTAL PARTICLES IN PLANTS

COMBINATIONS OF EXOGENOUS LIGHT FUNDAMENTAL PARTICLES, WITH PLANTS INDOGENOUS INTER ELECTRON NUCLEON PARTICLE COMPOUNDS: AND NEO- PARTICLE COMPOUND GENESIS INSIDE PLANT'S ELECTRONS AND NUCLEONS:

AIRBORN EXOGENOUS SUN-LIGHT FUNDAMENTAL PARTICLES (EX. - Y.- F.P.), SHINE DIRECT ON PLANTS ELECTRONS – NUCLEONS, THE EX. –Y. –F.P. INTER INTO PLANT'S INTERNAL ELECTRON- NUCLEON SUBSYSTEM-UNITS CHEMICAL LAB.S, THE EXOGENOUS LIGHT PARTICLES COMBINE WITH INOGENOUS INTER ELECTRONS NUCLEONS PARTICLE COMPOUND'S (F.P. – COMP.) AND PRODUCE LIGHT FUNDAMENTAL PARTICLE COMPOUNDS (Y. - F.P. _ COMP.), THESE PRODUCED LIGHT PARTICLE COMPOUNDS CONSTRUCT PLANTS INTERNAL ELECTRON - NUCLEON PARTICLE COMPOUND CONSTRUCTIONS THROUGH A. S. I. F.P. Mol. C.I.C.

THIS PHENOMENON IS NEO- PARTICLE COMPOUND GENESIS, IT IS NEO-ELECTRON, NEO– NUCLEON –GENESIS PHENOMENON AS WELL.

DEPENDING WHAT COLOR EXOGENOUS LIGHT PARTICLES COMBINE WITH PLANTS INTERNAL ELECTRON NUCLEON PARTICLE COMPOUNDS, UNDER A. S. I. F.P. Mol. C.I.C., THE PRODUCED INTERNAL ELECTRON-NUCLEON LIGHT PARTICLE – COMPOUNDS COLORS WILL BE EXACTLY THE SAME COLOR AS THE COMBINING EXOGENOUS LIGHT PARTICLES COLORS, IF COMBINING EXOGENOUS LIGHT PARTICLE IS GREEN, THE PARTICLE COMPOUNDS COLORS WILL TINT GREEN AS WELL, THE RED COLOR OR YELLOW COLOR PETALS COMBINING LIGHT FUNDAMENTAL PARTICLES COLORS, RESPECTEDLY ARE RED – LIGHT PARTICLES AND YELLOW LIGHT PARTICLES COLORS.

THIS PHENOMENON IS THE CAUSE FOR PLANTS ELECTRONS, NEUTRONS, CELLS, ATOMS, COLOR CHANGES AFTER THE COMBINATIONS WITH EXOGENOUS FUNDAMENTAL PARTICLES, THE PLANT CELLS TINT WITH COMBINING LIGHT PARTICLES COLOR, THE PARTICLE COMPOUND COLORS CHANGES ALSO INTO COMBINING LIGHT PARTICLES COLORS.

THIS IS PHENOMENON OF COLOR – GENESIS IN PLANTS, IF COMBINING LIGHT PARTICLE IS GREEN COLOR THEREFORE PLANT ELECTRON-NUCLEONS PARTICLE- COMPOUNDS CELLS GETS GREEN COLOR AS WELL, (OR COLOR- GENESIS PHENOMENON IN PLANTS).

EARTH'S SENSIBLE LIGHT FUNDAMENTAL PARTICLES

MAJORITIES OF SUN- ORIGIN BIOFRIENDLY LIGHT FUNDAMENTAL PARTICLES CLASSES, ARE NON SENSIBLE AND NOT - DISCOVERED LIGHT PARTICLES. BIOFRIENDLY NON- SENSIBLE LIGHT PARTICLE CLASSES CONSTRUCT MOST OF PLANET EARTH'S INTER ELECTRONS - NUCLEONS PARTICLE COMPOUND CONSTRUCTIONS, ALSO INTERNAL MONO- ATOMS STRUCTURES, AND NANO-UNIT CONSTRUCTIONS, NANO- UNIT CLOUDS, AND ATOM – CLOUDS.

EARTH'S DISCOVERED SENSIBLE LIGHT FUNDAMENTAL PARTICLES CLASSES, PRESENTLY ARE KNOWN AS LIGHT- WAVES, THE SENSIBLE LIGHT FUNDAMENTAL PARTICLES AT EARTH ARE: GREEN –COLOR LIGHT PARTICLES, RED – LIGHT PARTICLES, AND YELLOW COLOR LIGHT PARTICLE CLASSES, FROM OTHER SENSIBLEE PARTICLES ARE, BLUE COLOR FUNDAMENTAL PARTICLES, OR VIOLET COLOR LIGHT PARTICLE CLASSES, WHICH ALL OF THESE PARTICLES ALSO CONSTRUCT PLANET EARTHS ATOM'S LECTRON – NUCLEON PARTICLE COMPOUND CONSTRUCTIONS.

PARTICLE BIOCHEMISTRY

THE COLOR OF PLANT'S PARTICLE –COMPOUNDS, WHEN PLANT'S ELECTRON-NUCLEON COMPOUNDS NOT COMBINED WITH LIGHT-PARTICLES.

PLANT'S COLOR IS WHITE CLEAR WATER COLOR, WHEN PLANT HAS NO LIGHT PARTICLE COMBINATIONS

IN CASES WHEN PLANTS TISSUE'S, POSSESS NO CHEMICAL COMBINATIONS WITH AIRBORN LIGHT- FUNDAMENTAL PARTICLES, MOSTLY IN THESE CASES, THE CELL'S COLORS, THE ATOM'S COLOR, THE PLANT'S TISSUES COLOR ALL STAY COLORLESS, OR WATER- CLEAR WHITE COLOR, IN MOST TISSUES WHICH, THESE CELLS DO NOT POSSESS, OR MINIMALLY POSSESS LIGHT PARTICLES COMBINATIONS WITH INTER ELECTRON – NUCLEON PARTICLE COMPOUNDS, AND THERE IS ONLY MINIMAL OR, LESS LIGHT PARTICLE COMPOUNDS INSIDE THEIR PLANT'S ELECTRONS NUCLEONS.

THESE CASES MOSTLY OCCUR AT EARLY STAGES, WHEN IN THIGHTLY CLOSED THICK FLOWER BUDS, OR INSIDE THICK NIDUS, WHEN THE LIGHT FUNDAMENTAL PARTICLES ARE UNABLE TRANSIT INTO INSIDE UN-PENETRABLE CLOSED PLANT SPACES.

AT EARLY GROWTH STAGE OF THIGHTLY CLOSED THICK BUDS, WHICH DOES NOT ALLOW THE LIGHT PARTICLES INTER INTO THIGHTLY CLOSED THICK BUDS, AND DOES NOT ALLOW LIGHT PARTICLES CHEMICAL COMBINATIONS WITH INTER ELECTRONS NUCLEONS PARTICLE – COMPOUNDS TO TAKE PLACE, THIS PHENOMENON ALSO OCCUR INSIDE HARD SHELLS OF NUTS AT EARLY STAGE GROWTH, WHICH LIGHT PARTICLES IS UNABLE CROSS THICK SHELLS AND COMBINE WITH PIT CELL'S ELECTRONS NUCLEONS, THESE TISSUES STAY WHITE COLORLESS CLEAR BECAUSE TISSUES

ELECTRONS NUCLEONS POSSESS NO COMBINATIONS WITH LIGHT PARTICLES.

PARTICLE BIOCHEMISTRY

PLANT ATOMS –GENESIS, AND PLANTS CELLS NEO – GENESIS, BY GREEN LIGHT PATICLES

PLANT'S ELECTRON – NUCLEON PARTICLE COMPOUND – GENESIS, WITH GREEN LIGHT PARTICLE COMBINATIONS:

IN EARLY SPRING, WHEN THE EARLY PLANT STEM CELLS MULTIPLY WITH EXPONENTIAL SPEEDS, AND IN MANY SPECIES THE EARLY FLOWER'S THICK BUDS ARE THIGHTLY CLOSED, THE LIGHT PARTICLES ARE NOT CAPABLE FOR DIRECT INTERANCES INTO INSIDE CLOSED CUSPIDS, PETALS, OVARIES, STAMEN STAY COLORLESS.

ONLY THE CLOSED BUDS CUSPIDS, UNDER DIRECT DOMINANT Y. g. – F.P. INTERY INTO INSIDE CUSPID'S INTER ELECTRONS NUCLEONS SUBSYSTEM UNITS CHEMICAL LAB.S, HAVE BEEN COMBINED WITH INTER CUSPIDS INTERNAL ELECTRON-NUCLEON PARTICLE COMPOUNDS, AND HAVE PRODUCED GREEN LIGHT PARTICLE COMPOUNDS, AND THE Y. g. – F.P. COMPOUNDS HAVE BEEN CONSTRUCTED, THE CUSPIND'S ELECTRONS AND NUCLEONS MOLECULAR CONSTRUCTIONS WITH GREEN COLOR LIGHT PARTICLE COMPOUNDS, THROUGH THE A. S. I. F.P. Mol. C.I.C.

IN FIRST PHASE THE EXOGENOUS AIRBORN LIGHT

FUNDAMENTAL PARTICLES, EXCEPT THE DOMINENT GREEN LIGHT FUNDAMENTAL PARTICLES ARE NOT CAPABLE TO TRANS – CROSS THE CLOSED CUSPIDS THICK WALLS, AND THIS CUSPID CLOSURE IS THE BEST PROTECTION PROCEDURE FOR OVARIES, STAMEN, PETALS OF THE EARLY STAGE FLOWER'S STEM CELLS.

AT THIS STAGE ONLY DOMINANT GREEN LIGHT PARTICLES CURRENTS, ARE CAPABLE DIRECTLY TO TRANS-CROSS THICK CUSPIDS, THROUGH P.A.S. –PARTICLE CIRCULATION SYSTEMS INTO INSIDE BUD'S CLOSED CAVITY, AND COMBINE WITH INTERNAL ELECTRON-NUCLEON PARTICLE COMPOUNDS, AND PRODUCE GREEN LIGHT PARTICLES –COMPOUNDS CONSTRUCTIONS INSIDE ELECTRON'S- NUCLEON'S OF PETALS, OVARIES, STAMENS, WHEN THE BUDS, CUSPIDS ARE CLOSED. AT THIS STAGE, THE PETALS, STAMENS, OVARIES PARTICLE COMPOUND CONSTRUCTIONS MOSTLY ARE CONSTRUCTED WITH GREEN LIGHT PARTICLE COMPOUNDS, ENTIRE FLOWERS CUSPIDS, PETALS, OVARIES, AND STAMEN ALL ARE GREEN COLOR.

THIS IS PHENOMENON OF PARTICLE COMPOUND NEO-GENESIS, THE ELECTRON NEO- GENESIS, THE NUCLEON NEO -GENESIS PHENOMENON.

AT THIS STAGE THE OTHER LIGHT FUNDAMENTAL PARTICLES ARE RECESSIVE, THEY ARE NOT CAPABLE TRANS – CROSS THROUGH P.A.S. CURRENTS CIRCULATION SYSTEMS, INTO INSIDE CAVITY OF THIGHTLY CLOSED, THICK CLOSED CUSPIDS, (ONLY DOMINANT Y. g. –F.P. POSSESS THIS CAPABILITY).

PARTICLE BIOCHEMISTRY
ALTERNATING STABLE STEM CELL'S MUTATIONS
WITH MEDIA CHEMICAL FORMULARY CHANGES

CONSTRUCTION OF ANIMALS AND PLANTS IN-UTERO FETUC (FETUS –GENESIS PHENOMENON) UNDER

ALTERNATING MUTATION SEQUENCES (AMS), WHEN ONE STABLE MUTATION ALTERNATE WITH ANOTHER STABLE MEDIAL CHIMICAL FORMULARY CHANGES, IN UTERO AUTONOMOUSLY SEQUENTIALLY WITH EXPONENTIAL SPEEDS OF MULTIPLICATIONS AND PRODUCTIONS BETWEEN THE STEM CELLS, WHEN RAPIDLY ONE STABLE MUTATION STEM CELLS MUTATE INTO ANOTHER STABLE STEM CELL WITH EXPONENTIAL SPEED OF INTERACTIONS UNDER A. S. I. N.-U. –C.I.C.

ALTERNATING SEQUENTIAL STEM CELL MUTATIONS: THE MUTATIONS SEQUENTIALLY ALTERNATING WITH STEM CELLS MEDIA CHEMICAL FORMULARY CHANGES, IS CAUSE OF NEW TISSUE-GENESIS (MUTATIONS), IT IS CAUSE OF NEO-ORGAN - GENESIS, NEO-SYSTEM GENESIS, AND NEW FETUS CONSTRUCTION IN ANIMALS AND PLANT EMBRYONIC

STATES. THIS IS PHENOMENON OF COMPLETE FETUS –GENESIS IN EMBRYO, THROUGH AUTONOMOUS SEQUENTIAL ALTERNATING MUTATION SEQUENCES, WHICH THEIR ALTERNATE WITH M.C.F.C., AND PRODUCE DIFFERENT TISSUES AND CELLS (CAUSE CELL DIFFERENTIATIONS), THIS IS THE CAUSE HOW ANIMALS AND PLANTS BODY CONSTRUCT A WHOLE CREATURE.

IN PLANTS THE MEDIA C.F.C. (CHANGING ONE COMBINING LIGHT PARTICLES TO OTHER) UNDER EVOLUTION PATH, CAUSE PLANTS STEM CELLS MUTATE AND CHANGE TO OTHER KIND STEM CELLS. THE MEDIA CHEMICAL FORMULARY CHANGES (M.C.F.C. = Y.F.P CHANGE), CAUSE STEM CELLS TRANSFORMS (MUTATIONS) THROUGH NEO PARTICLE COMPOUND GENESIS PHENOMENON FROM ONE TO OTHER, UNDER EVOLUTION ORDERS.

OPPOSITE OF ABOVE INTERACTIONS ALSO IS TRUE, THE CREATED NEW STEM CELLS MUTATION PRODUCE M. C. F. C., ONE FOLLOW OTHER IN CYCLES.

---

1 - STEM CELL'S MUTATIONS CAUSE, CHEMICAL FORMULARY CHANGES OF STEM CELL'S MEDIA (OR: M. C. F. C.) BECAUSE OF CELL'S METABOLISM.

2- M. C. F. C. (CHANGING MEDIA LIGHT PARTICLE IN AIR) CAUSE STEM CELL MUTATIONS (MUTATION OF EARLY STAGE

FAST GROWING STEM CELLS). IN EMBRYO, THE GREEN COLOR LIGHT PARTICLE WAS ONLY COMBINING LIGHT PARTICLES, FOLLOWING OPENING CUSPIDS ALL OTHER LIGHT PARTICLES ALSO DIRECTLY INTER INTO SUBSYSTEM UNITS LAB.S AND COMBINE WITH PARTICLE COMPOUND, CHANGE THE CELL'S COLOR.

3- AT OPENING CUSPIDS, THE FLOWERS EXPOSE TO DIRECT OTHER COLOR LIGHT PARTICLES, DIFFERENT LIGHT PARTICLES INTER AND COMBINE WITH FLOWERS INTER ELECTRON NUCLEON PARTICLE COMPOUNDS, PRODUCE PARTICLE COMPOUND CHANGE IN FLOWER TISSUES (MUTATION GENESIS).

THIS IS MUTATION OF STEM CELLS FROM ONE LIGHT PARTICLE COMPOUND, TO ANOTHER PARTICLE COMPOUND. THIS PHENPMENON IS NEO- STEM CELL GENESIS UNDER A. S. I. F.P. Mol. C.I.C., ALSO BECAUSE OF PHENOMENON OF ALTERNATING CELL'S MEDIA CHEMICAL FORMULARY CHANGES WHICH ALTERNATE WITH SEQUENTIAL STEM CELL NEO-MUTATIONS. THIS ALTERNATIONS ARE MAIN CAUSES OF CELL DIFFERENTIATIONS IN EMBRYO AND CONSTRUCTIONS OF WHOLE PLANT TISSUES AT EMBRYONIC PLANT STATES.

4- IN ANIMALS AND HUMAN EMBRYO THE SEQUENCES OF STEM CELLS MUTATIONS ALTERNATIONS WITH MEDIA CHEMICAL FORMULARY CHANGES ALTERNATION IS THE MOST IMPORTANT STEM CELLS DIFFERENTIATION FACTORS, AND MAIN CAUSE FOR DIFFERENT TISSUE-GENESIS, DIFFERENT ORGAN-GENESIS, DIFFERENT SYSTEM GENESIS AND

DIFFERENT ANIMAL TOTAL BODY GENESIS DURING THE IN-UTERO FETUS DEVELOPMENTS AND GROW TO COMPLETE WHOLE WELL CONSTRUCTED NEW ANIMAL BABY READY TO BE DELIVERED TO OUTSIDE, ABOVE PHENOMENON IS ANIMAL GENESIS PHENOMENON INSIDE UTERUS ABOVE IS ONE EXAMPLE FROM CELL-BASE AND BIOMOLECULAR BASE MEDIA – C.F.C. WHEN IT IS ALTERNATING WITH STEM CELLS ALTERNATION PHENOMENONS.

PARTICLE BIOCHEMISTRY

STEM CELLS DIFFERENTIATIONS, OR STEM CELL'S ALTERNATING STABLE MUTATION SEQUENCES AND PLANTS GENESIS (IN EARLY STAGE PLANT GROWTH PERIODS)

IN EARLY STAGE PLANT ATOM-CELL MUTATIONS, ONLY ARE UNDER A.S. I. F.P. Mol. C.I.C., AND MEDIA CHEMICAL FURMULARY CHANGES IS CHANGING INTERNAL ELECTRON NUCLEON PARTICLE COMPOUNDS, THESE CHANGES SECONDARILY PRODUCE ATOMS, CELL'S AND TISSUES CHANGES TO OTHER COLORS STRUCTURES (MUTATIONS), SUCH AS CHANGING COLORLESS CELLS TO GREEN CELLS, OR CHANGING GREEN CELLS TO DIFFERENT OTHER COLOR CELLS, THROUGH USE OF M.C.F.C., AS WELL THROUGH A.S. I. F.P. Mol. C.I.C.,

GREEN LIGHT PARTICLES COMBINATIONS UNDER A.S. I. F.P. Mol. C.I.C., AT FIRST SEQUENCE THEY PRODUCE GREEN COLOR MUTANT ATOMS, CELLS, TISSUES, AND PLANTS, IN ORDER TO CHANGE FIRST SEQUENCE MUTANTS, INTO OTHER COLORS LIGHT PARTICLES MUTANT STEM CELL'S PRODUCTIONS,

IN PLANTS UNDER A. S. I. F.P. Mol. C.I.C. COMBINED WITH AIRBORN OTHER COLOR LIGHT PARTICLES, THEIR INTERNAL CELL'S ATOMS PARTICLE- COMPOUND CONSTRUCTIONS EXPLANIED IN THIS BOOK. BUT CELL BASE AND BIOMOLECULAR BASE MUTATIONS IN ANIMAL EMBRYO IS SUBJECT OF OTHER NEXT VOLUMES.

ATOM MUTATIONS, CELL MUTATIONS, AND PLANT MUTATIONS

THE AIRBORN, Y. g. – F.P. INTER INTO P.A.S. CURRENTS, AND FROM THERE THROUGH ELECTRON NUCLEONS P.C.S., INTER INTO INSIDE CLOSED BUDS, DIRECTLY INTER INTO INSIDE PETALS, OVARIES, STAMENS, INTERNAL ELECTRONS –NUCLEONS CHEMICAL LAB.S, AND COMBINE WITH INTERNAL ELECTRONS NUCLEONS PARTICLE COMPOUNDS, THROUGH REGENERATIVE A. S. I. F.P. Mol. C.I.C., PRODUCE THE GREEN COLOR LIGHT PARTICLE COMPOUNDS OF ENTIRE PETALS, OVARIES, STAMENS, OF EARLY STAGE FLOWERS BUDS, WHEN THE FLOWER CUSPIDS ARE THIGHTLY CLOSED, THROUGH REGENERATIVE A. S. I. F. P. Mol. C.I.C.THE CONSTRUCTED GREEN COLOR LIGHT PARTICLE COMPOUNDS CONSTRUCT MOST PARTICLE MOLECULAR CONSTRUCTIONS OF THE PETALS, OVARIES, STAMENS, ALL WITH GREEN COMPOUNDS, WHEN STILL THE CUSPIDS ARE CLOSED, THE ENTIRE CLOSED BUDS IN MOST SPECIES, AT EARLY GROWTH STAGES, CHANGE INTO GREEN COLOR CELLS AND TISSUES AND PETALS, CUSPIDS, OVARIES AND STAMENS, THIS IS PRIMARY SEQUENTIAL ATOM-CELL-TISSUE MUTATIONS IN PLANTS.

THE GREEN LIGHT FUNDAMENTAL PARTICLES, THROUGH THE SAME PATTERNS IN EARLY STAGE PLANT GROWTH, INTER INTO INTERNAL ELECTRONS, NUCLEONS OF FRUITS, NIDUS, ETC. AND COMBINE WITH FRUITS, PITS, INTERNAL ELECTRONS NUCLEONS PARTICLE COMPOUNDS, AND PRODUCE GREEN COLOR LIGHT PARTICLE _ COMPOUNDS, AND CONSTRUCT GREEN COLOR ATOMS, CELLS, BECAUSE THE CONSTRUCTED ELECTRONS AND NUCLEONS COLORS ARE GREEN THROUGH THE GREEN COLOR LIGHT PARTICLE COMPOUNDS, WHICH ALL OF THOSE STRUCTURES ARE IN THE SAME COLORS OF COMBINING GREEN LIGHTS PARTICLES.

THIS IS THE REASON, ALL EARLY STAGE, PETALS, OVARIES, STAMENS, FRUITS, LEAVES, STEMS, ETC. ALL ARE GREEN COLOR, BECAUSE THEIR CONSTRUCTING PARTICLE COMPOUNDS ARE GREEN –LIGHT PARTICLE COMPOUNDS, WHICH THE CELLS AND TISSUES ADDITIONALLY TINT GREEN BECAUSE OF GREEN LIGHT PARTICLE COMPOUND CONSTRUCTIONS ALL TISSUES ARE GREEN COLORS.

THIS PHENOMENON IS PLANT CELL NEO –GENESIS, AND TISSUE NEO –GENESIS PHENOMENON OF PLANTS. IT IS A STABLE ATON MUTATION, STABLE CELL MUTATION, STABLE PLANTS ORGANS AND SYSTEMS MUTATIONS THROUGH THE ALTERNATING MEDIA CHEMICAL FORMULARY STRUCTURES THE CONSTRUCTIONS OF INSIDE ATOMS AND CELLS CHANGE INTO ANOTHER PARTICLE COMPOUNDS, THIS IS ALSO PARTICLE BASE CHEMICAL INTERACTIONS WHICH CAUSING THE MUTATIONS.

THE PARTICLE BIOCHEMISTRY

ALTERNATIONS M. C. F. C. FOR FETAL CELLS IN ANIMAL EMBRYONIC STAGES, PRODUCE FETAL EMBRYONIC STEM CELL'S ALTERNATING MUTATIONS, AND VICE VERSA

THESE MUTATIONS SEQUENCES IN STEM CELLS BY ITSELF, CAUSE SECONDARY STEM CELL MEDIA CHEMICAL FORMULARY CHANGES (M. C.F.C.) FOR EMBRYONIC STEM CELLS, UNDER CLOSED CYCLES OF ABOVE, ONE STABLE STEM CELL MUTATION CYCLE FOLLOWS, ONE M. C.F.C. SEQUENCE, AND ONE M. C.F.C. CAUSE STEM CELL'S MUTATIONS SECONDARILY, ONE SEQUENCE FOLLOW THE OTHER SEQUENTIALLY.

IN ANIMAL SPECIES, THE IN -UTERO FETUS, THROUGH EXACT COPY OF FUNDAMENTAL PARTICLE MUTATIONS OF PLANTS, BUT IN ANIMAL EMBRYO, UNDER SAME PATTERNS, THE BIOMOLECULES AND STEM CELLS MUTATIONS FROM ONE TO OTHER SEQUENTIALLY OCCUR, AND AT THE SAME TIME EACH SEQUENCE NEW MUTATION CAUSE, THE MEDIA CHEMICAL FORMULARY CHANGES AROUND THE PRODUCED MUTANT STEM CELL, AND THE NEWLY PRODUCED NE CONSTRUCTION MEDIA CHEMICAL FORMULARY SECONDARILY CAUSE STEM CELLS MUTATION TO OTHER, THIS CYCLES CIRCULATE IN CLOSED CHAINS AND CYCLES, ONE AFTER THE OTHER, UNIT THE TOTAL CONSTRUCTIONS OF A NORMAL FETAL ORGANS, SYSTEMS, TISSUES AND CELLS ACHIEVED AND READIED FOR DELIVERIES.

IN THE PLANTS SPECIES, BECAUSE OF PLANTS ELECTRONS-NUCLEONS PARTICLE COMPOUNDS COMBINATIONS WITH GREEN DOMINENT LIGHT PARTICLES, THE LEAVES, STEMS, CUSPIDS, PETALS, STAMEN, OVARY, FRUITS, AND ALL PLANTS AT EARLY STAGE, ALL ARE GREEN COLORS. THIS IS PHENOMENON OF GREEN COLOR PLANT TISSUE –GENESIS, IT IS STABLE MUTATION AND IS CAPABLE ALTERNATE TO OTHER COLOR TISSUES.

THIS TISSUE NEO-GENESIS PHENOMENON LATER ALTERNATE, AND CHANGE COLOR OF FLOWERS, FRUITS TO OTHER COLORS, WHEN THE OTHER COLOR LIGHT FUNDAMENTAL PARTICLS AFTER OPENING BUDS, DIRECTLY INTER INTO PETALS, STAMEN, OVARY AND COMBINE WITH THEIR PARTICLE COMPOUNDS, THESE SECONDARY A. S. I. F.P. Mol. C.I.C. CAUSE CHANGES OF FLOWER COLORS, FRUITS INTO OTHER COLORS, THIS IS STABLE MUTATION OR STABLE TRANSFORMATION OF PLANT TISSUES, THESE ARE PHENOMENON OF STABLE MUTATIONS, ONE AFTER THE OTHER, IN PATHS OF MOLECULAR EVOLUTIONS.

THE FETUS –GENESIS PHENOMENON (EMBRYO- GENESIS PHENOMENON) IN ANIMALS AND PLANTS,

PHENOMENON OF AUTONOMOUS SEQUENTIAL MUTATIONS (ASM) ALTERNATING SECODARY TO, MEDIA CHEMICAL FORMULARY CHANGES (M. C. F. C.) AND VICE VERSA:

1 - MEDIA CHEMICAL FORMULARY CHANGES (M. C. F. C.) CAUSE, THE STEM CELLS MUTATIONS (S. C. M.).

2 - THE STEM CELLS MUTATIONS (S. C. M.) CAUSE, THE MEDIA CHEMICAL FORMULARY CHANGE (M. C. F. C.).

3 – THE M. C. F. C. ALTERNATE WITH EMBRYONIC CELL MUTATIONS. DURING EMBRYONIC INTERNAL UTERUS GROWTH, IS THE CAUSE FOR CELL DIFFERENTIATIONS IN ANIMALS,

EACH M. C. F. C. CREATE ANOTHER MUTATION, THE PRODUCED MUTANT CELL'S METABULISM SECONDARILY PRODUCE M. C. F. C. AGAIN IN CYCLE,

CLOSED CYCLES CHAINS OF MUTATIONS ALTERNATING WITH M. C. F. C. ONE AFTER THE OTHER DURING EMBRYONIC PHASES OF FETUS, CAUSE CREATIONS OF DIFFERENT TISSUES , ORGANS, BODY SYSTEMS, IN FETUS AND AFTER OTHER, AT THE END CONSTRUCT A FULL SIZE BABY WELL DEVELOPED AT THE BIRTH, THIS PHENOMENON IS: ANIMAL –GENESIS AND PLANT GENESIS PHENOMENON IN EMBRYO, DEPENDING IT IS HAPPENING IN EARLY STAGE PLANTS STEM CELLS GROWTH, OR IT IS TAKING PLACE INSIDE ANIMAL EARLY STAGE STEM CELLS GROWTH, INSIDE THE UTERUS.

THIS IS THE PHENOMENON OF: MUTATIONS ALTERNATING WITH, M. C.F. C. INSIDE UTERUS, AND THIS IS THE CAUSE OF FETUS –GENESIS IN ANIMALS

PLANT GENESIS PHENOMENON

FUNDAMENTAL PARTICLE BIOCHEMISTRY

IN STEM CELLS RAPID GROWTHS PHASES IN EARLY SPRING, WHEN FLOWER BUDS OPENS, IMMEDIATELY PETALS, STAMEN, OVARY ARE UNDER DIRECT EXPOSURE OF ALL DIFFERENT COLOR LIGHT FUNDAMENTAL PARTICLES, AT THIS PHASE ALL LIGHT PARTICLES SUCH AS, RED, YELLOW, BLUE, VIOLET, COLOR LIGHT PARTICLES DIRECTLY AIRBORN INTER INTO INSIDE PETALS, OVARIES, STAMENS, ELECTRONS – NUCLEONS SUBSYSTEM-UNITS CHEMICAL LAB.S, AND COMBINE DIRECT WITH INTER ELECTRON –NUCLEON PARTICLE COMPOUNDS, AND PRODUCE LIGHT PARTICLE COMPOUNDS WHICH THOSE NEWLY PRODUCED LIGHT PARTICLE COMPOUNDS COLORS, ARE EXACTLY THE SAME COLOR OF COMBINING INCIDENT EXOGENOUS LIGHT PARTICLE COLOR, THAT COMBINED WITH INTERNAL ELECTRON NUCLEON PARTICLE COMPOUNDS.

PRODUCED FLOWER ATOMS, CELLS, TISSUES ARE IN EXACT COLORS OF INCIDENT EXOGENOUS COMBINING LIGHT FUNDAMENTAL PARTICLE'S COLORS. THESE PRODUCED NEW PARTICLE COMPOUNDS, WHICH HAS DIFFERENT COLORS FLOWER CONSTRUCTIONS, AND PLANT'S INTERNAL ELECTRON- NUCLEON PARTICLE COMPOUND CONSTRUCTIONS ARE DIFFERENT COLORS, THIS IS LIGHT PARTICLE –COMPOUND GENESIS PHENOMENONS, NEO - ELECTRON- NUCLEON GENESIS PHENOMENONS, ETC.

AS EXPLAINED BEFORE AT FIRST SEQUENCES A.S. I. F.P. Mol.

C.I.C. IN EARLY STAGES PLANTS GROWTH THERE IS GREEN LIGHT PARTICLES COMBINATIONS DOMINANCE WITH MOST OF FLOWERS FRUITS AND PLANTS ELECTRONS AND NUCLEONS, AT EARLY STAGES STEM CELL RAPID GROWTHS PLANTS STAGES NOT ONLY ATOM BASE AUTONOMOUS SEQUENTIAL INTERACTIONS AND STEM CELL BASE SEQUENTIAL CHEMICAL INTERACTIONS TAKE PLACE AT EXPONENTIAL SPEED RATES, AT THE SAME TIME SUN LIGHT FUNDAMENTAL BASE A. S. I. F.P. Mol. C.I.C. ALSO ALL RUNING WITH EXPONENTIAL SPEEDS OF CHEMICAL INTERACTIONS, AND CONSTRUCTING DIFFERENT PLANTS TISSUES INTERNAL ELECTRON NUCLEON CONSTRUCTIONS, THROUGH EXPONENTIAL SPEEDS IN ALL CELL- ATOM- PARTICLE BASE CHEMICAL SEQUENCES WITH COOPERATIONS AND COORDINATIONS OF EACH OTHER, IN SHORT TIME EMBRYONIC EARLY SPRING STEM CELL STAGE, AT EARLY SPRING EMBRYONIC STEM CELL EXPONENTIAL GROWTH PERIOD GREEN LIGHT PARTICLE COMBINE WITH ALL PLANTS TISSUES AT OVER 85 % OF EXISTING PLANTS SPECIES.

UNDER SECOND SEQUENCES OF A. S. I. F.P. Mol. C.I.C. IN LATER STAGES, WHEN FLOWERS OPENS AND THERE IS DIRECT EXPOSURES OF FLOWER ATOMS TO ALL LIGHT PARTICLES, AT THIS STAGE THE PRE-CONSTRUCTED GREEN COLOR LIGHT PARTICLE COMPOUNDS UNDER DEGENERATIVE A. S. I. F.P. Mol. C.I.C.AT INSIDE FLOWERS FRUITS PLANTS PETALS-OVARIES-STAMEN, AS WELL AS AT FRUITS ETC. ALL BREAK DOWN INTO SMALLER CONSTRUCTING MOLECULAR STRUCTURES, SUCH AS FREE GREEN LIGHT PARTICLES, WHICH GREEN LIGHT PARTICLE ALSO RELEASE AS FREE AIRBORN

PARTICLES SCAPE TO AIR AND DISAPPEAR.

IN SECOND SEQUENCES THGERAFTER, THE PLANTS FLOWERS, FRUITS, PARTICLE COMPOUNDS, ETC. THEY ARE NOT GREEN COLOR ANY MORE, AND THEY HAVE BEEN COMBINED WITH OTHER COLOR LIGHT PARTICLES, AND HAVE CONSTRUCTED THE ELECTRONS NUCLEONS CONSTRUCTIONS WITH ANOTHER COLOR LIGHT PARTICLE COMPOUDS, AND FLOWERS, PETALS, STAMENS, FRUITS, ETC. HAVE BEEN CHANGEED COLORS TO THE ANOTHER ONES, SUCH AS RED, YELLOW, VIOLET, BLUE COLORS, DEPENDING WHICH COLOR LIGHT PARTICLE COMBINED WITH WHICH PART ATOMS OF THE PLANTS LEAVES, FLOWERS, FRUITS, STEMS, ETC. HAVE THE CHEMICAL COMBINATIONS, UNDER THE A.S. I. F.P. Mol. C.I.C.

WHEN CUSPIDS AND FLOWERS OPEN IN LATER STAGES, THERE AFTER OTHER COLOR LIGHT PARTICLE SUCH AS RED, YELLOW, BLUE, VIOLET, ETC. THROUGH DIRECT SUN SHINE, INTER INTO FLOWERS FRUITS PLANTS INTER ELECTRONS NUCLEON SUBSYSTEM – UNITS CHEMICAL LABS. AND UNDER A. S. I. F.P. Mol. C.I.C. COMBINE WITH INTER ELECTRONS NUCLEONS PARTICLE COMPOUNDS AND PRODUCE DIFFERENT COLORS LIGHT PARTICLE COMPOUNDS OF RED, YELLOW, BLUE, VIOLET, OR OTHER COLOR PARTICLE COMPOUNDS, PARTICLE COMPOUNDS COLORS ARE THE SAME AS COMBINING LIGHT PARTICLES, ALSO THE ATOMS, CELLS COLORS ALSO ARE THE SAME AS THEIR CONSTRUCTING PARTICLE COMPOUNDS.

DEPENDING WHICH COLOR LIGHT PARTICLE COMBINING TO FRUITS, FLOWERS, PLANTS PARTICLE COMPOUNDS, THEY PRODUCE THAT COLOR LIGHT PARTICLE COMPOUNDS INSIDE PLANTS FLOWERS, FRUITS, ELECTRONS, NUCLEONS, ATOMS, THE COLOR OF PLANT CHANGE INTO COMBINING LIGHT PARTICLES COLORS.

THIS PHENOMENON IS PLANT- GENESIS PHENOMENON, AND PHENOMENON OF TRANSFORMATIONS OF ONE COLOR PLANT TISSUES TO THE OTHER COLORS PLANTS, THESE PHENOMENONS ALL ARE STABLE MUTATIONS WHICH ARE TAKING PLACE UNDER PREVIOUS EVOLUTION ORDERS OF STABLE MUTATIONS, SEQUENCES, ONE AFTER OTHER.

PRODUCTION OF AIRBORN LIGHT PARTICLE CLOUDS, OR PARTICLE CLOUD GENESIS

VISIBLE PARTICLE CLOUDS GENESIS

COMBINING AIRBORN EXOGENOUS LIGHT PARTICLES, WITH INDIGENOUS LIGHT PARTICLE COMPOUNDS

H – O – H ATOMS ARE CONSTRUCTED FROM LIGHT FUNDAMENTAL PARTICLE - COMPOUND CONSTRUCTIONS, THE H – O – H OR WATER ATOMS INTERNAL ELECTRON, NUCLEON PARTICLE COMPOUNDS HAVE BEEN CONSTRUCTED FROM LARGE QUANTITIES OF RED, YELLOW, GREEN, BLUE, VIOLET, LIGHT PARTICLE COMPOUNDS CONSTRUCTIONS.

THE INTERNAL ELECTRON, NUCLEON INDIGENOUS

DIFFERENT LIGHT PARTICLE COMPOUNDS CONSTRUCTIONS OF WATER ATOMS, UNDER THE A. S. I. F.P. Mol. C.I.C. COMBINE WITH EXTERNAL ELECTRON- NUCLEON EXOGENOUS PARTICLES, SUCH AS WHITE LIGHT PARTICLES, AFTER COMBINATION THE RELEASED, FREE LIGHT PARTICLES IN FORMS OF AIRBORN PARTICLE CLOUDS SCAPE INTO AIR, AS RED, YELLOW, GREEN, BLUE, VIOLET DIFFERENT COLOR LIGHT PARTICLES CLOUDS AGAIN.

PARTICLE CLOUD GENESIS:

IN WARM TEMPERATURE SUMMER CONDITIONS, SHINE WHITE COLOR LIGHT FUNDAMENTAL PARTICLES AIRBORN INTO ELECTRONS, NUCLEONS OF VAPORIZED H – O – H AIRBORN ATOMS AT AIR. AND COMBINE WHITE INCOMING EXOGENOUS LIGHT PARTICLES FROM AIR WITH INTERNAL ELECTRON NUCLEON DIFFERENT LIGHT PARTICLE COMPOUNDS OF DIFFERENT COLORS, SUCH AS: RED LIGHT PARTICLE COMPOUNDS, YELLOW AND GREEN LIGHT PARTICLE COMPOUNDS, OR COMBINE WITH INTERNAL ELECTRON NUCEON BLUE AND VIOLET DIFFERENT LIGHT PARTICLE COMPOUNDS, UNDER REGENERATIVE A. S. I. F.P. Mol. C.I.C.

THESE MULTI- A.S.I. P.F. Mol. C.I.C. BETWEEN EXOGENOUS WHITE LIGHT PARTICLE COMBINATION WITH INDIGENOUS DIFFERENT PARTICLE COMPOUNDS COMBINATIONS, WILL PRODUCE DIFFERENT COLOR LIGHT PARTICLES, WHICH THESE RELEASED LIGHT PARTICLES SCAPE INTERNAL ELECTRON, NUCLEON REGIONS BACK TO AIR, THE PRODUCED PARTICLE CLOUD AT AIR ARE VISIBLE AS: RED LIGHT PARTICLE

CLOUDS. YELLOW LIGHT PARTICLE CLOUD, GREEN PARTICLE CLOUDS, BLUE AND VIOLET LIGHT PARTICLE CLOUDS. AND DECIPATE AIRBORN TO SPACE SHORTLY AFTER PRODUCTION AND DISAPPEAR.

FROM OTHER A. S. I. F.P. Mol. C.I.C. OCCURRENCES IN COMBINATIONS OF AIRBORN BIOFRIENDLY WHITE COMPOSITE LIGHT PARTICLES, WHICH THESE EXOGENOUS WHITE LIGHT PARTICLE HAVE BEEN COMPOSED FROM VARIOUS DIFFERENT CLASSES OF DIFFERENT COLOR LIGHT FUNDAMENTAL PARTICLES, AND DURING THE A. S. I. F.P. Mol. C.I.C. SOME OF THESE WHITE FUNDAMENTAL PARTICLES COMPOSITE COMPONENTS, DIRECTLY INTER INTO DIFFERENT H – O – H INTERNAL ELECTRON NUCLEON SUBSYSTEM UNITS CHEMICAL LAB.S AND COMBINE WITH INTERNAL ELECTRON NUCLEON PARTICLE COMPOUNDS OF H – O- H ATOMS, AND THE OTHER REMAINING LIGHT FINADAMENTAL PARTICLES HAS TO SCAPE INTERNAL ATOM SPACE, TO THE OUT SIDE INTO AIR, AND THESE FREE MULTI- CLASS PARTICLES ARE VISIBLE AS VISIBLE LIGHT PARTICLE-CLOUDS, STANDING WITH EACH OTHER UNDER SPECIFIC PARTICLES LAWS AND RULES. WHICH WE CAN SEE THEM AS VISIBLE PARTICLE CLOUDS IN AIR.THIS IS PHENOMENON OF LIGHT PARTICLE CLOUD – GENESIS, UNDER A. S. I. F.P. Mol. C.I.C.

ALL A. S. I. F.P. Mol. C.I.C. ARE FAST WITH EXPONENTIAL SPEEDS RATE OF CHEMICAL INTERACTIONS, ABOVE AUTONOMOUS SEQUENTIAL CHEMICAL INTERACTIONS LASO ARE WITH NO EXCEPTIONS, THE ABOVE A.S.I. F.P. Mol. C.I.C. WILL IMMEDIATELY CEASE, IF WE STOP H-O-H VEPORIZATIONS,

ALSO IT WILL STOP, IF WE CEASE THE SHINING OF EXOGENOUS LIGHT PARTICLES, BOTH WILL STOP THE PRODUCTIONS OF PARTICLE CLOUDS SIMILAR TO "ON – OFF" ELECTRIC KEY INSTANTLY. THIS IS SIMPLE TEST TO DEMONSTRATE EXISTING PARTICLE CLOUD VISIBLE TO NAKED EYES. THIS PHENOMENON IS GENESIS OF LIGHT FUNDAMENTAL PARTICLE CLOUDS, THROUGH A.S.I. F.P. Mol. C.I.C.

PRODUCTION OF, VISIBLE LIGHT FUNDAMENTAL PARTICLE –CLOUDS, IN NATURE, HIGH IN SKY

THIS PHENOMENON IS A NATURAL OCCURRING A. S. I. F.P. Mol. C.I.C. TAKING PLACE HIGH IN SKY, BETWEEN COMBINATION OF AIRBORN, EXOGENOUS, SUN ORIGIN, INCOMING, WHITE COLOR LIGHT FUNDAMENTAL PARTICLE WHICH COMBINING WITH H – O – H INTERNAL ELECTRON NUCLEON, MULTIPLE DIFFERENT LIGHT PARTICLE COMPOUNDS, WHICH THESE A. S. I. F.P. Mol. C.I.C. TAKING PLACE HIGH IN CLOUDS OF THE SKY, (CLOUDS ARE H. – O – H. ATOMS), THESE A. S. I. F.P. Mol. C.I.C. PRODUCE MULTIPLE DIFFERENT COLOR LIGHT PARTICLE CLOUDS OF DIFFERENT COLORS SUCH AS: RED, YELLOW, GREEN, BLUE AND VIOLET COLORS LIGHT FUNDAMENTAL PARTICLE CLOUDS, HIGH IN THE SKY. WHICH THESE MULTI COLOR LIGHT PARTICLE CLOUDS ARE VISIBLE FROM MILES AWAY FROM EARTH, AND THESE AIRBORN LIGHT PARTICLE CLOUDS ARE CALLED RAIN BOW BY PUBLIC. RAIN BOW PRODUCTION IS EXACTLY THE SAME A. S. I. F.P. Mol. C.I.C. AS EXPLAINED IN PREVIOUS PAGES AND ABOVE.

PHENOMENON OF PARTICLE – CLOUD GENESIS

GENESIS OF AIRBORN PARTICLE CLOUDS

COMBINING EXOGENOUS LIGHT PARTICLES WITH CRYSTALINE INTERNAL ELECTRONS-NUCLEONS INDIGENOUS LIGHT-PARTICLE COMPOUNDS:

CRYSTALINE ELECTRONS NUCLEONS ARE LIGHT-PARTICLE CONSTRUCTED STRUCTURES AT MOST, THE WATER (H – O – H ATOMS), AND CRYSTALINE ATOMS, BOTH ARE LIGHT FUNDAMENTAL PARTICLE CONSTRUCTED ATOMS, AND THEIR INTERNAL ELECTRON-NUCLEON CONSTRUCTIONS HAVE BEEN CONTRUCTED FROM DIFFERENT KIND OF LIGHT FUNDAMENTAL PARTICLE COMPOUNDS, THE DIFFERENT NON VISIBLE LIGHT PARTICLE CLASSES, AND DIFFERENT VISIBLE LIGHT FUNDAMENTAL PARTICLE COMPOUNDS SUCH AS RED LIGHT PARTICLE COMPOUNDS, GREEN LIGHT PARTICLE COMPOUNDS, BLUE OR VIOLET COLOR LIGHT PARTICLE COMPOUNDS OR YELLOW COLOR LIGHT PARTICLE COMPOUNDS CONSTRUCT DIFFERENT ELECTRON-NUCLEOM PARTICLE COMPOUND CONSTRUCTIONS OF CRYSTALINE ATOMS,

IN A TEST, SHINE AIRBORN EXOGENOUS WHITE LIGHT FUNDAMENTAL PARTICLES INTO CRYSTALINE INTERNAL ELECTRON NUCLEON SUBSYSTEM CHEMICAL LAB.S, AND COMBINE EXOGENOUS WHITE LIGHT PARTICLES WITH INDIGENOUS INTERNAL ELECTRON NUCLEON DIFFERENT COLOR LIGHT PARTICLE COMPOUNDS, THROUGH A. S. I. F.P. Mol. C.I.C., THE CHEMICAL INTERACTIONS BETWEEN

EXOGENOUS INCOMING WHITE LIGHT FUNDAMENTAL PARTICLES, AND INDIGENOUS CRYSTALINE INTER ELECTRON-NUCLEON DIFFERENT COLOR INTER ELECTRON –NUCLEON LIGHT PARTICLE COMPOUNDS WILL PRODUCE DIFFERENT COLOR RELEASED FREE LIGHT FUNDAMENTAL PARTICLES, AS WELL AS WHITE- LIGHT –PARTICLES –COMPOUNDS, AS PARTICLE BY PRODUCTS.

THROUGH THESE CHEMICAL INTERACTIONS, THE INTERACTION PRODUCED AND RELEASED DIFFERENT COLOR FREE LIGHT FUNDAMENTAL PARTICLE CLOUDS, OF DIFFERENT COLORS SUCH AS RED, YELLOW, GREEN, BLUE, VIOLET, ETC. COLOR LIGHT FUNDAMENTAL PARTICLE CLOUDS, WILL BE VISIBLE AT THE EXIT SIDE OF CRYSTALINE, AS MULTI COLOR DIFFERENT KINDS LIGHT PARTICLE CLOUDS, WHICH EACH COLOR IS REPRESENTATIVE OF ONE CLASS LIGHT FUNDAMENTAL PARTICLE CLOUD, ARE DETECTABLE IN PARTICLE CLOUD THROUGH NAKED EYES, SHORTLY AFTER PRODUCTION WILL DECIPATE INTO AIR, AS FREE PARTICLES DISAPPEAR IN SPACE.

THESE PRODUCED FREE AIRBORN DIFFERENT COLOR LIGHT FUNDAMENTAL PARTICLE CLOUDS, AND EACH COLOR PARTICLE CLOUD IS REPRESENTAVIES OF THEIR COLOR SPECIFIC CLASS LIGHT PARTICLE CLASSES. THESE PARTICLE CLASS PARTICLE CLOUDS WILL STAND ORDERLY UNDER PARTICLE ORDERS WITH EACH OTHER, SUCH AS: RED, YELLOW, GREEN, BLUE, VIOLET COLOR LIGHT PARTICLE CLOUDS IN SPECTRAL LINE, AND AFTER TERMINATIONS AND CEASE OF A. S. I. F.P. Mol. C.I.C., THE INDIVIDUALLY PRODUCED LIGHT

PARTICLE-CLOUDS OF DIFFERENT COLORS, WILL DECIPATE IN STANTLY AND TRAVEL IN SPACE TOWARD IT'S DESTINATIONS. FOLLOWING ARE SOME DEMONSTRATIONS FROM INTERACTIONS:

Y. w. – F.P. + Y. r. – F.P. _ COMP. -----------→ Y. w. – F.P. _ COMP. + Y. r. –F.P.

←--------------

Y. w. – F.P. + Y. y. –F.P. _ COMP. ------------------→ Y. w. – F.P. _ COMP. + Y. y. – F.P.

←------------------

Y. w. – F.P. + Y. g. –F.P. _ COMP. -------------------------→ Y. w. F.P. _ COMP. + Y. g. F.P.

←-----------------------------

Y. w. – F.P. + Y. b. – F.P. _ COMP. ----------------------→ Y. w. – F.P. _ COMP. + Y. b. –F.P.

←----------------------

Y. w. – F.P. + Y. v. – F.P. _COMP. ------------------------→ Y. w. – F.P. _ COMP. + Y. v. –F.P.

←----------------------

COMMUNICATING OF ATOMS WITH EACH OTHER STORAGE OF ENVIRONMENTAL KNOWLEDGE OF UNIVERSE IN INFORMATION IMAGE PARTICLE CLOUDS FORMS, INSIDE ELECTRONS NUCLEONS. EXCHANGE OF THESE F.P.-I.I.- P. cl. BETWEEN DIFFERENT ATOMS:

BIOFRIENDLY LIGHT, SONIC, ELECTRIC, THERMAL FUNDAMENTAL PARTICLE INFORMATION IMAGE PARTICLE CLOUDS (Y. S. E. T. – F.P. – I.I. – p.cl.) ARE THE MOST COMMON TYPES PARTICLE CLOUDS, TRANSIT AND EXCHANGE INTELLIGENCE IN FORMS OF PARTICLE CLOUDS INFORMATIONS AND IMAGES, BETWEEN DONOR AND RECIPIENTS ELECTRONS NUCLEON, AND INTERACT WITH EACH OTHER UNDER A. S. I F.P. Mol. C.I.C. IN THE EARTH, AND SIMILARLY IN ENTIRE UNIVERSE BETWEEN THE ATOMS.

THIS PHENOMENON IS EXCHANGES OF EXISTING INTELLIGENCES AND KNOWLEDGES OF PLANETS AND UNIVERSES, IN FUNDAMENTAL PARTICLE CLOUDS INFORMATIONS IMAGES FORMS, AND THIS PHENOMENON IS COMMUNICATIONS OF EXISTING PLANETARY AND UNIVERSAL INTELLIGENCES AND KNOWLEDGES, IN THE FORMS OF Y. S. T. E. – F.P. – I.I. – P. cl., BETWEEN THE MOST OF EXISTING ELECTRONS, NUCLEONS.

IN PERFORMING THESE NANO- TASKS, SOME ELECTRONS, NUCLEONS, AND NANO- UNITS ARE MUCH INTELLIGENCE, AND CAPABLE THAN THE OTHERS TO PERFORM THESE INTER NANO-UNIT COMMUNICATIONS AND STORAGES BETTER THAN OTHERS. EVEN SOME OF THE NANO – UNITS ARE ENTIRELY ILLITERATE IN THESE ESSENTIAL UNIVERSAL TASKS, AND CAN NOT COMMUNICATE AND STORE ANY INFORMATION INAGES CLOUDS AT ALL.

COMBINING (STORING) S.Y.-F.P.-I.I.-P.cl.(VIRTUAL CLOUDS OF OUTSIDE) INSIDE ELECTRONS NUCLEONS: RECORDING GENESIS OF OUTSIDE WORLD'S INFORMATION IMAGE ENCYCLOPEDIA INSIDE CNS ATOMS: PHENOMENON OF PSYCHE-GENESIS AND THOUGHT GENESIS:

EXISTING SUBJECTS AT EARTH, EMIT PARTICLE CLOUDS COPIES FROM THEMSELVES, WHICH THESE PARTICLE CLOUDS COPIES HAS BEEN MADE MOSTLY FROM LIGHT AND SONIC FUNDAMENTAL PARTICLE CLOUDS (S. Y. – F.P. – I.I. – P. cl.) IN PLANET EARTH, THESE PARTICLE CONSTRUCTED CLOUDS ARE IN EXACT FORMS OF ORIGINAL SUBJECTS, BUT THESE SUBJECT'S - COPY CLOUDS ALL MADE FROM FUNDAMENTAL PARTICLES IN CLOUD FORMS. THESE PARTICLE CLOUDS COPIES OF EXISTING SUBJECTS ARE CAPABLE TO MOVE, SPEAK, PERFORM, WALK, STUDY AND DO ANY THINGS THAT ORIGINAL SUBJECTS IS CAPABLE TO DO EXACTLY THE SAME, AS ORIGINAL SUBJECT'S CAPABILITIES.

THE CLOUDS ARE INFORMATIONS, IMAGES, PERFORMANCES IN CLOUD FORMS, MADE FROM DIFFERENT PARTICLES. THE CLOUDS ALL MADE FROM FUNDAMENTAL PARTICLES, WHEN THESE CLOUDS (S. Y. –F.P. – I.I. – P. cl.) RECORDS AND STORES INSIDE CNS ELECTRONS AND NUCLEONS IN PARTICLE CLOUD COMPOUND FORMS AND CONSTRUCT THE CNS ELECTRONS NUCLEONS PARTICLECOMPOUND CONSTRUCTIONS, THESE STORED A. S. Y.- F.P. – I.I. – P. cl. ARE EXACTLY SIMILAR TO ORIGINAL SUBJECTS PERFORMANECES INSIDE THE ELECTRONS AND NUCLEONS OF CNS. CLOUDS PERFOMANCES INSIDE CNS ELECTRONS NUCLEONS ARE EXACTLY THE SAME AS ORIGINAL SUBJECTS PERFORMING INSIDE BRAIN. THIS IS THE WHAT THE ANIMALS FEEL INSIDE THEIR BRAINS WHEN THEY ARE THINKING. (WHAT THEY ARE SEEING INSIDE THEIR BRAINS, ALL ARE PARTICLE CLOUD PERFORMANCES, AND THERE IS NO REAL SUBJECT INSIDE ELECTRONS AND NUCELONS OF THE CNS).

IN ALL PLANETS, THE ALL FUNDAMENTAL PARTICLE INFORMATION IMAGE CLOUDS, ONLY ARE MADE FROM NATIVE FUNDAMENTAL PARTICLES, THOSE ARE THE FUNDAMENTAL PARTICLES SPECIFIC AND NATIVE FOR THOSE PLANETS.

PARTICLE CLOUD DONERS

AT EARTH MOST COMMON INFORMATION AND IMAGES PARTICLE CLOUDS ARE MADE FROM BIOFRIENDLY: LIGHT PARTICLES, SONIC PARTICLES, ELECTRIC PARTICLES, THERMAL FUNDAMENTAL PARTICLE CLOUDS (Y. S. E. T. – F.P. – I.I. – p. cl.), THE MOST OF SUBJECTS AND EXISTING THINGS OF PLANETH EARTH ARE CAPABLE TO PRODUCE AND EMIT Y. S. T. E. – F.P. – I.I. – p.cl. INTO AIR OR TRANSIT THROUGH OTHER FUNDAMENTAL PARTICLE TRANSITION ROUTES, THE PARTICLE CLOUD DONERS (F.P. – I.I. – p. cl. – DONERS) ARE THOSE SUBJECTS THAT, THEY EMIT AND PRODUCE FROM THEMSELVES F.P. – I.I.- P. cl. - COPY INTO SPACE OR AIR.

THE EXISTING LIVING THINGS, SENSARY ORGANS ELECTRONS-NUCLEONS POSSESS CAPABILITIES UNDER ATTRACTIVE GRAVITON FORCE TO ATTRACT PARTICLE CLOUDS FROM AIR, AND CAPTURE THE PARTICLE CLOUDS FROM AIR AS RECIPIENTS OF INFORMATION IMAGES PARTICLE CLOUDS FROM ANY SOURCE. THIS PHENOMENON IS COMMUNICATIONS BETWEEN DIFFERENT INTELLIGENCE SYSTEMS, THE ATTRACTED PARTICLE-CLOUDS THROUGH NANO NEURAL FIBERS TRANSMIT INTO CNS ELECTRONS NUCLEONS, COMBINE WITH CNS SENSARY CENTERS PARTICLE COMPOUNDS,

AND CONSTRUCT THE INTER ELECTRON NUCLEONS PARTICLE COMPOUND CONSTRUCTIONS.

THIS IS PHENOMENON OF INTELLIGENCE EXCHAGE BETWEEN ATOMS ELECTRONS NUCLEONS, THE A. S. I. F.P. Mol. C.I.C. POSSESS EXTREME RESPONSIBILITIES OF RECORDING, INTERACTIONS, AND RETRIEVAL PHENOMENON OF THE INFORMATIONS AND IMAGES PARTICLE CLOUDS INSIDE ATOMS, OR TO OUTSIDE ELECTRONS, NUCLEONS.

PARTICLE CLOUD (Y. S. E. T. – F.P. – I.I. – p. cl.)
CURRENTS CIRCULATIONS SYSTEMS

BETWEEN PARTICLE DONERS AND RECIPIENTS

THE PRODUCED AND EMITTED PARTICLE CLOUDS (Y. S. E. T. – F.P. –I.I. – P. cl.) FROM DONER'S SUBJECTS TRAVEL AIRBORN IN AIR AT ANY DIRECTION, UNTIL THE RECIPIENT'S SUBJECTS SENSARY ELECTRONS-NUCEONS, THROUGH THEIR ATTRACTIVE FORCES OF GRAVITON ATTRACT PARTICLE CLOUDS INTO INSIDE RECIPIENT ELECTRONS- NUCLEONS SUBSYSTEM-UNITS CHEMICAL LAB.S, THE INCOMING EXOGENOUS INFORMATION IMAGE PARTICLE CLOUDS COMBINE WITH PARTICLE COMPOUNDS OF RECIPIENT ELECTRONS NUCLEON, AND PRODUCE INFORMATION-IMAGE PARTICLE CLOUD - COMPOUNDS AND BECOME PART OF CONSTRUCTIONS OF RECIPIENT ATOMS. (S. Y. T. E. – F.P. – I.I. – p. cl.- COMP.) THIS IS PHENOMENON OF STORAGE OF ENVIRONMENTAL SUBJECTS INFORMATION IMAGES INSIDE RECIPIENT ATOMS, IN PARTICLE CLOUD COMPOUND FORMS,

STORING SCIENCE AND KNOWLEDGE
INSIDE ELECTRONS-NUCLEONS
INTELLIGENCE – GENESIS OF ELECTRONS, NUCLEONS

AIRBORN PARTICLE CLOUD CIRCULATION SYSTEMS
BETWEEN ANIMALS AND OTHER EXISTING THINGS

THROUGH AIRBORN PARTICLE –CLOUDS (S. Y. – F.P. – I.I. – P. cl.) CIRCULATION SYSTEMS, BETWEEN DIFFERENT INDIVIDUALS, EITHER THEY ARE PARTICLE CLOUD EMITTERS, AND OTHER INDIVIDUALS ARE PARTICLE CLOUD RECEIVERS, THROUGH AIRBORN CLOSED CYCLE PARTICLE CLOUD CIRCULATION SYSTEMS BETWEEN DIFFERENT CNS ELECTRONS AND NUCLEONS DURING A COMMUNICATION, SOME INDIVIDUALS EMIT PARTICLE CLOUDS, AND OTHERS RECEIVES PARTICLE CLOUDS, THESE F.P. – I.I.- P. cl. CIRCULATE INSIDE A AIRBORN CLOSED CIRCUIT OF PARTICLE CLOUD CIRCULATION SYSTEMS BETWEEN INDIVIDUALS CNS ELECTRONS NUCLEONS.

THE DIFFERENT INDIVIDUALS THROUGH EXCHANGES OF S. Y. –I.I.- P. cl. BETWEEN EACH OTHER, THEY ANALIZE THE PARTICLE CLOUDS THROUGH A.S. I. F.P. Mol. C.I.C. INSIDE CNS ELECTRONS NUCLEONSAND CONSTRUCT THE PARTICLE CLOUD RESPONSES.

DIFFERENT SUBJECTS EMIT PARTICLE –CLOUDS – COPY OF SELF INTO SPACE, THE PRODUCED SONIC- LIGHT FUNDAMENTAL PARTICLE INFORMATION IMAGE PARTICLE CLOUDS FROM A GIVEN SUBJECT (S. Y. – I.I. - F.P. –P.cl.) TRAVEL

AIRBORN OR THROUGH OTHER PARTICLE TRANSMISSION ROUTES FROM DONER SUBJECT TOWARD RECIPIENT ATOMS.

THE RECIPIENT ATOMS THROUGH THEIR ATTRACTIVE GRAVITON FORCES ATTRACT INCOMING S. Y. – F.P. – I.I. – P. cl. INTO P. A. S. AND THEREAFTER PARTICLE – CLOUDS FROM P.A.S INTER INTO INSIDE RECIPIENT ELECTRONS- NUCLEONS SUBSYSTEM –UNITS AND COMBINE WITH INTER ELECTRONS NUCLEONS DIFFERENT PARTICLE COMPOUNDS AND PRODUCE S. Y. – F.P. – I.I. – P. cl. _ COMPOUNDS AND BUILD RECIPIENTS ATOMS INTERNAL ELECTRON-NUCLEON PARTICLE COMPOUND CONSTRUCTIONS, AND STORE S.Y. – F.P. –I.I. – P. cl. INSIDE RECIPIENT ELECTRONS – NUCLEONS IN PARTICLE – CLOUD _ COMPOUND FORMS, THIS IS PHENOMENON OF STRORAGE OF DIFFERENT ENVIRONMENTAL SUBJECTS INFORMATION – IMAGE PARTICLE – CLOUDS INSIDE ATOMS, THIS PHENOMENON IS ALSO STORAGE OF INFORMATION-IMAGE OR KNOWLEDGE AND SCIENCE INSIDE ELECTRONS – NUCLEONS, WHICH UNDER DEGENERATIVE A. S. I. F.P. Mol. C.I.C. CAN BE RETRIEVED AS FREE PARTICLE-CLOUDS SCAPE TO OUT OF ELECTRONS –NUCLEONS INTO SPACE AGAIN.

PARTICLE-CLOUDS CAN TRANSIT THROUGH PERIPHERAL ATOM SPACE OF SOLIDS, LIQUIDS, PLASMA FROM ONE LOCATION TO OTHER THROUGH PRODUCED PARTILE-CURRENTS, ATOM-CLOUDS ARE DIFFERENT SUBJECTS, BUT FOLLOW THE SAME PARTICLE BASE RULES,

S. Y. - F. P. - I. I. – P.cl. FROM PERIPHERAL ATOM-SPACE (P.A.S.)

INTER INTO PERIPHERON, TRAVEL THROUGH PERIPHERON-NUCLEUS FUNDAMENTAL-PARTICLE CIRCULATION SYSTEMS (P. C. S.) PARTICLE CURRENTS IN PERIPHERON AND NUCLEUS DIVID INTO SMALLER PATICLE CIRCULATION BRANCHES, PARTICLE CLOUDS FINALLY FROM SIDE BRANCHES, INTER INTO DIFFERENT ELECTRONS PROTONS AND NEUTRONS SUBSYSTEMS-UNITS CHEMICAL -LAB.S, AND COMBINE WITH DIFFERENT ELECTRON-NUCLEON PARTICLE COMPOUNDS AND PRODUCE NEW S. Y. –F.P. – I.I. – pl. _ COMPOUNDS CONSTRUCTING, CONSTRUCT THE NEW ELECTRON-NUCLEON PARTICLE COMPOUND CONSTRUCTIONS, THAT POSSESS SPECIFIC INFORMATIONS KNOWLEDGE IN REGARD TO SPECIFC OUTSIDE FACTS, INFORMATIONS IMAGES ARE STORED INTACT INSIDE ATOMS (PHENOMENON OF INTELLIGENCE ELECTRONS-NUCLEONS GENESIS), THIS IS RECORDING OF INFORMATIONS – IMAGES OF OUTSIDE FACTS INSIDE RECIPIENTS ELECTRON – NUCLEON SUBSYSTEM –UNITS IN FORMS OF PARTICLE – CLOUD-COMPOUND CONSTRUCTIONS.

THOUGHT CURRENTS GENESIS PHENOMENON

PSYCHE – GENESIS PHENOMENON

PHENOMENONS OF DECISION MAKING, MEMORIZATION AND ORDER ISSUING PHENOMENONS BY CNS CENTERS ELECTRONS –NUCLEONS:

IN LIVING THINGS INSIDE CNS ELECTRONS NUCLEONS THROUGH A. S. I. F.P. Mol. C.I.C. THE PRE-EXISTING

CONSTRUCTED INDIGENOUS S. Y. –F.P. –I.I.- P. cl. _ COMPOUNDS COMBINE WITH INCOMING EXOGENOUS Y. S. – F.P. – I.I. – P. cl. AND PRODUCE DIFFERENT PARTICLE BY PRODUCTS, PARTICLE CLOUD COMPOUNDS, THE DIFFERENT PARTICLE CLOUDS CIRCULATE BETWEEN DIFFERENT CNS CENTERS THROUGH, CNS PARTICLE CIRCULATING SYSTEMS, BETWEEN DIFFERENT CNS CENTERS ELECTRONS NUCLEONS, THE PARTICLE CLOUDS INTERACT WITH DIFFERENT OTHER CNS CENTERS INTER ELECTRON- NUCLEON PARTICLE CLOUD COMPOUNDS, ARE IN CONTINUEOUS COMPLEX A. S. I. F.P. Mol. C.I.C., ALL OF THESE PARTICLE CLOUD INTERACTIONS AND CIRCULARTION OF PARTICLE CLOUD CURRENTS INSIDE CNS. PARTICLE CIRCULATION SYSTEM, THE PHENOMENONS BIOLOGICALLY SENSE BY LIVING THINGS AS THOUGHT-CURRENTS AND PSYCHE PHENOMENON.

IN EARTH THE LIVING THINGS LIGHT- SOUND SENSER ORGANS ELECTRONS – NUCLEONS CAPTURE INCOMING EXOGENOUS Y. S. – F.P. – I.I. – P. cl. OF ENVIRONMENTAL EXISTING THINGS PARTICLE CLOUDS, THROUGH ATTRACTIVE GRAVITON FORCES THE PARTICLE CLOUDS INTER INTO ELECTRONS NUCLEONS CHEMICAL LAB.S, AND INCIDENT PARTICLE CLOUDS CHEMICALLY COMBINE WITH NTER ELECTRONS NUCLEON PARTICLE COMPOUNDS. PRODUCE PARTICLE CLOUD COMPOUNDS. AND THERE AFTER THROUGH NEURAL PARTICLE TRANSIT – ROUTES, AND PARTICLE CLOUD CIRCULATION SYSTEMS, THE EXOGENOUS INFORMATION IMAGE PARTICLE-CLOUDS CARRIED INTO DIFFERENT CNS-CENTERS ELECTRONS- NUCLEONS SUBSYSTEM UNITS CHEMICAL LAB.S, AND COMBINED WITH PARTICLE COMPOUNDS,

BUILD CNS SENSARY CENTERS ELECTRONS- NUCLEONS PARTICLE COMPOUND CONSTRUCTIONS, WHICH THE PARTICLE CLOUD COMPOUNDS ARE S. Y. –F.P. – I.I. – p. cl. –STORAGE SYSTEMS.

THE S. Y. - F.P. –I.I. – P.cl. ARE KNOWLEDGE OF OUTSIDE EVENT'S INFORMATION AND IMAGES IN FORMS OF PARTICLE CLOUDS, THAT ARE STORED IN PARTICLE COMPOUND FORMS, AT INSIDE CNS SENSARY CENTERS ELECTRONS NUCLEONS, THESE PARTICLE CLOUD STORAGES ARE ENVIRONMENTAL KNOWLEDGE AND SCIENCE STORING SYSTEMS, INSIDE CNS ELECTRONS AND NUCLEONS IN FORMS OF S. Y. – F.P. –I.I.- p.cl. _COMPOUNDS, WHICH THESE STORAGES ARE THE BUILDING BLOCKS OF LEARNING, SCHOOLING, TEACHING, EDUCATIONS, TRAININGS, ETC.

THROUGH RECORDING, STORING OF OUTSIDE WORLD S. Y. – F.P. – I.I. – p. cl. IN PARTICLE CLOUD- COMPOUNDS FORMS INSIDE SENSARY CNS CENTERS ELECTRONS-NUCLEONS, THESE KNOWLEDGE STOCK PILES AND INTELLIGENCE STORAGE SYSTEMS PHENOMENONS OF OUTSIDE WORLD EVENTS INFORMATIONS AND IMAGES AT INSIDE CNS ELECTRONS AND NUCLEONS, ARE PHENOMENON OF ACCUMULATIONS OF SCIENCE AND KNOWLEDGE OF OUTSIDE INFORMATION, IN PARTICLE-CLOUD COMPOUND FORMS, AT INSIDE CNS ATOMS, THIS PHENOMENON IS BUILDING BLOCKS OF LEARNING, TRAINING, TEACHING, EDUCATIONS, AT HOMES, SCHOOLS, UNIVERSITY AS WELL AT ALL SOCIAL, RELIGIOUS, POLITICAL, AND EDUCATIONAL FIELDS.

THE A. S. I. F.P. Mol. C.I.C. BETWEEN THESE INDIGENOUS S. Y. – F.P. – I.I. – p. cl. _ COMPOUNDS WITH OTHER INCOMING EXOGENOUS INCOMING S. Y. – F.P. – I.I. – p. cl. INSIDE CNS ELECTRON- NUCLEON SUBSYSTEM – UNITS CHEMICAL LAB.S, AND INTERACTIONS OF ALL OF THESE WITH ENTIRE CNS DIFFERENT CENTER ELECTRONS –NUCLEON POPULATIONS, WITH EXPONENTIAL SPEEDS INTERACT WITH EACH OTHER, PRODUCE PARTICLE CLOUDS AND PARTICLE CLOUD CURRENTS, AND THEIR CHEMICAL INTERACTIONS WITH EACH OTHER CAUSE PRODUCTIONS OF NEW PARTICLE BY PRODUCTS, WHICH ALL OF THESE EVENTS BIOLOGICALLY SENSED AS PHENOMENON OF PSYCHE AND THOUGHT CURRENTS BY LIVING THINGS, THIS IS CAUSE OF PSYCHE-GENESIS, THOUGHT - GENESIS AND THOUGHT CURRENT – GENESIS AT LIVING THINGS.

THESE NEWLY PRODUCED PARTICLE- CLOUD –COMPOUNDS, CONSTRUCT DIFFERENT CNS ELECTRONS, NUCLEONS, PARTICLE COMPOUND CONSTRUCTIONS, UNDER REGENERATIVE – DEGENERATIVE A. S. I. F.P. Mol. C. I. C., MOST OF INCOMING EXOGENOUS S. Y. – F.P. I.I. – p. cl. INTERACT WITH PRE-EXISTING INTERNAL ELECTRON NUCLEON S. Y. – F.P. – I.I. – p. cl. _ COMPOUNDS IN DIFFERENT CNS CENTERS ELECTRONS NUCLEONS.

DURING MEMORIZATION AND DECISION MAKING PHENOMENONS, THE DIFFERENT A. S. I. F.P. Mol. C.I.C. TAKE PLACE EXPONENTIALLY BETWEEN INTERACTING EXOGENOULS INCIDENT PARTICLE CLOUD, WHEN THOSE INTERACTING AND COMPARING, THESE NEWLY INTERED PARTICLE

CLOUDS, WITH THOSE PRE-EXISTING PARTICLE CLOUD COMPOUNDS, IN ORDER THROUGH THESE DIFFERENT COMPARISIONS, THE CNS ELECTRONS NUCLEONS FINALLY REACH INTO CONCLUSIONS, AND DECISION MAKINGS, WITH ENTIRE INDIGENOUS PREVIOUSLY CONSTRUCTED EXISTING INFORMATIONS IMAGES PARTICLE CLOUD COMPOUNDS, INSIDE ENTIRE ELECTRONS- NUCLEONS OF DIFFERENT CNS CENTERS.

FINDING BEST SOLUTIONS AND CHOICES THROUGH THESE INNORMOUS DIFFERENT A. S. I. F. P. Mol. C.I.C. THROUGH COMPARING ALL OF PREVIOUSLY RECORDED S. Y. –F.P. – I.I.- p.cl. COMPOUNDS IN DIFFERENT CNS –CENETRS ELECTRONS-NUCLEONS, FINALLY THE HIGHER DECISION MAKING BRAIN CENTERS ELECTRONS – NUCLEONS SELECT BEST SOLUTIONS THROUGH ABOVE RESEARCHES, AND EXECUTE THE PROPER DECISIONS AND RESPONSES FOR ONGOING SITUATION EVENTS, DECIDES, SELECTS, AND EXCUTE PROPER ORDERS ON BASIS OF FINAL FINDINGS, BETWEEN THE DIFFERENT RESEARCHES, BETWEEN THE EXISTING FACTS IN RECORDED INFORMATIONS-IMAGE FORMS, WHICH EXIST IN FORMS OF PARTICLE-COMPOUNDS FORMS FOR COMPARISON OF ENTIRE BRAIN CENTERS, KNOWLEDGES INSIDE ELECTRONS-NUCLEONS, THESE ARE PEOCEDURE USED DURING DECISION MAKING PHENOMENONS, AND MEMORIZING PHENOMENONS, AND EXECUTION AND ORDER ISSUING PHENOMENONS INSIDE DIFFERENT CENTER CNS ELECTRONS –NUCLEONS.

PARTICLE CLOUD TRANSMISSION ROUTES
AIRBORN PARTICLE CLOUD CIRCULATION SYSTEMS
CIRCUITS BETWEEN DIFFERENT INDIVIDUALS
INTERNAL BODY PARTICLE CLOUD CIRCULATION SYSTEMS

TRANSMISSION ROUTES OF S. Y. E. T. – F.P. – I. I. – P. cl.:

PARTICLES AND PARTICLE CLOUDS POSSESSES TWO MAIN PARTICLE CIRCULATION SYSTEMS: 1- INTERNAL BODY PARTICLE CIRCULATION SYSTEMS, 2 – EXTERNAL BODY PARTICLE CLOUD CIRCULATION SYSTEMS.

1 – EXTERNAL BODY PARTICLE CLOUD CIRCULATION SYSTEMS:

THE PARTICLE CLOUDS MOSTLY TRANSMIT AIRBORN FROM ONE INDIVIDUAL TO OTHER IN CLOSED PARTICLE CLOUD CIRCULATION SYSTEMS (PCS). IN SONIC, ELECTRIC AND LIGHT PARTICLE CLOUDS, WHEN TRANSMISSION BETWEEN MULTIPLE INDIVIDUAL TAKE PLACE, ALMOST ALL S. Y. –F.P. – I.I. – P. cl. TRANSIT AIRBORN IN CLOSED PARTICLE CLOUD CIRCULATION PATHS AT MOST.

THE INTER-INDIVIDUAL LIVING THINGS PARTICLE TRANSMISSIONS ARE AIRBORN. SOME TIMES THE PARTICLE CLOUDS TAKE DIFFERENT TRANSMISSION ROUTES, MOSTLY INTER INTO BODY THROUGH SENSARY FUNDAMENTAL PARTICLE TRANSIT ROUTES. ALSO THERE IS TRANSMISSION THROUGH FOOD PRODUCTS AS WELL, EVEN THERMAL PARTICLES,

ELECTRIC PARTICLES QUICKLY TRANSIT THROUGH SKIN AS WELL. IN PLANTS MOST PARTICLES TRANSIT AIRBORN, TRANSIT THROUGH ROOTS ALSO HAS POSSIBLITIES.

IN EARTH TRANSMISSION OF BIOFRIENDLY SONIC LIGHT FUNDAMENTAL PARTICLE INFORMATION IMAGE PARTICLE CLOUDS (S. Y. – F.P. –I.I. – P. cl.), BETWEEN HUMANS AND ANIMALS MOSTLY TRANSIT AIRBORN BETWEEN DIFFERENT INDIVIDUALS, THE PROCESS IS PARTICLE CLOUD TRANSMISSION SYSTEMS IN CLOSED AIRBORN CIRCULATIONS BETWEEN DIFFERENT INDIVIDUALS, AND PARTICLE CLOUDS AIRBORN TRAVEL BACK AND FORTH BETWEEN HUMAN SPECIES, BOTH NORMAL PARTICLE CLOUDS AS WELL AS ABNORMAL PARTICLE CLOUDS ALL TRANSMIT AIRBORN FROM ONE HUMAN TO OTHER IN AIRBORN PARTICLE CLOUD CIRCULATION SYSTEM CIRCUITS.

2- INTERNAL BODY PARTICLE AND PARTICLE CLOUD CIRCULATION SYSTEMS:

THE MAIN PARTICLE CIRCULATION SYSTEMS INSIDE ANIMAL BODY, CIRCULATE INSIDE INTERNAL NEURAL NANO-DUCT'S PARTICLE CLOUDS CIRCULATIONS SYSTEMS. NEURAL SYSTEMS EITHER IN NANO- LEVELS, OR IN MICRO LEVELS, AND MACRO LEVELS, ALL CIRCULATE INSIDE VOLUNTARY AND NON- VOLUNTARY, CENTRAL AND PERIPHERAL NERVOUS SYSTEMS, PARTICLE CIRCULATION SYSTEMS (PCS).

THE TRANSIT OF S. Y. – F.P. - I.I. - p. cl. INSIDE ANIMALS BODY MOSTLY TRANSIT THROUGH BI-DIRECTIONAL FINE NEURAL

FIBERS EFFERENT AND AFFERENT PARTICLE CURRENTS, WHICH CONNECTING DIFFERENT ALL HUMAN ORGANS TOTAL BODY ELECTRON-NUCLEONS PARTICLE POPULATIONS TO CNS TOTAL POPULATION ELECTRON-NUCLEONS IN DIFFERENT CNS CENTERS, THIS IS BIDIRECTIONAL FUNDAMENTAL PARTICLE CURRENTS POPULATIONS CONNECT TO CNS ATOMS TOTAL POPULATIONS IN DIFFERENT CENTERS, THROUGH INTELLECTUAL PARTICLE CENTER SYSTEMS THE DIFFERENT CENTRAL AND PERIPHERAL ELECTRON NUCLEON POPULATIONS COORDINATE PHYSICAL CHEMICAL BIOLOGICAL FUNCTIONS WITH EACH OTHER, THROUGH PARTICLE PRECISION ACCURACIES.

CLASSIFICATION OF PARTICLE CLOUDS

THE S. Y. T. E. – F.P. – I.I. – P. cl. DIVIDE INTO TWO MAJOR CLASSES: 1- NORMAL PARTICLE CLOUDS. 2 – ABNOMRMAL PARTICLE CLOUDS.

THE PSYCHIATRIC DISEASES ARE PRODUCED THROUGH THE TRANSMISSION OF ABNORMAL S. Y. – F.P. – I. I. – P. cl. FROM ONE PARTICLE CLOUD INFLICTED INDIVIDUAL, TO RECIPIENT OTHER INDIVIDUALS. AND PSYCHIATRIC DISEASES ARE PARTICLE CLOUD (S. Y. – F.P. – I.I. –p. cl.) TRANSMITTED DISEASES.

THE PARTICLE CLOUD TRANSMITTED DISEASES

(PSYCHIATRIC DISORDERS ARE ABNORMAL-PARTICLE CLOUD TRANSMITTED DISEASES).

NEWBORN AT BIRTH, DO NOT POSSESSES ANY SIGNIFICANT AMOUNT S. L. T. E. – F.P. – I.I. – P. cl. – COMPOUNDS INSIDE CNS ELECTRONS AND NUCLEONS. THE PARTICLE CLOUDS TRANSMISSION FROM MOTHER'S P.C.S. TO FETUSES BRAIN ELECTRONS AND NUCLEONS, THROUGH FETAL P. C. S.

CONNECTIONS TO MOTHER IS NEGLIGIBLE AMOUNT WHEN THE BABY BORN. THE FETAL INTERNAL ELECTRONS NUCLEONS FUNDAMENTAL PARTICLES INFORMATION IMAGE CLOUD CONSTRUCTIONS INSIDE UTERUS IS NEGLIGIBLE, THE S. Y. E. T. – F.P. – I.I. – P.cl. INTERACTIONS, THE STORAGE OR RETRIEVAL PHENOMENONS MOSTLY START AFTER THE BIRTH, AND CONTINUE DURING ALL LIFE LONG.

DURING THE LIFE, ALL THE NORMAL OR ABNORMAL F.P. – I.I.- P.cl. TRANSMIT INTO INDIVIDUALS CNS ELECTRONS NUCLEONS AND CONSTRUCT THE BRAINS INTERNAL ELECTRON - NUCLEON PARTICLE COMPOUND CONSTRUCTIONS. THE NORMAL OR ABNORMAL PARTICLE CLOUD TRANSMISSIONS DURING LIFE AFTER THE BIRTH, PRODUCE AND CONSTRUCT EITHER NORMAL OR ABNORMAL INDIVIDUAL PARTICLE COMPOUND CONSTRUCTIONS AND DEVELOP TO ADVANCED PSYCHE-GENESIS.

ALL OF THE S. Y. E. T. – F.P. – P. cl. OF NORMAL PSYCHE OR ABNORMAL PSYCHE, BEHAVIOR DISORDERS, NEUROSIS, DEPRESSIVE OR ELATED DISORDERS, PSYCHOSIS, NEUROSIS, ANXIETY, COMPULSIVE DISORDERS, NORMAL OR ABNORMAL GENIUSES, DIFFERENT CRIMINAL BEHAVIORS AND CRIMINALS, ETC. ALL THESE DIFFERENT PARTICLE CLOUDS TRANSMIT THROUGH OUT THE LIFE PROGRESSIVELY FROM OTHER INDIVIDUALS AND ENVIRONMENTS INTO NEWBORN INDIVIDUALS CNS, AND CONSTRUCT PARTICLE COMPOUND CONSTRUCTIONS OF BRAINS OF CHILDREN THROUGH USE OF RECEIVED PARTICLE CLOUDS.

THE E. Y. S. T. – F.P. – I.I. – P. cl. TRANSMISSIONS FROM ONE INDIVIDUAL TO OTHER INDIVIDUAL CONSTRUCT THOUGHT CURRENT SYSTEMS WHICH ARE PARTICLE CLOUD CIRCULATION SYSTEMS, AND PSYCHE MOSTLY IS ACQUIRED PHENOMENON, CONGENITAL NEUROLOGICAL DISEASES ARE DIFFERENT ISSUES, ALTHOUGH THOSE ALSO MAY CAUSE MENTAL DISORDERS.

THE ABNORMAL FUNDAMENTAL PARTICLE TRANSMITTED DISEASES, MOSTLY ARE PROTRACTED AND CHRONIC TYPES DISORDERS, FOR BEING INFLICTED BY PARTICLE CLOUDS DISORDERS, MOSTLY REQUIRE PROTRACTED TRANSMISSION TIMES, BUT IN SOME CASES PARTICLE CLOUD TRANSMIT IS FAST AND QUICK, EVEN IN SOME CASES COUPLE EXPOSURES AND PARTICLE CLOUDS TRANSMIT MAY PRODUCE PARTICLE INFLICTIONS.

PSYCHOGENESIS – ABILITIES OF INDIVIDUALS ARE DIFFERENT FROM EACH OTHER, SOME INDIVIDUALS HAVE STRONG ABILITIES TO ALTER THE PSYCHE OF OTHERS. ALSO THERE ARE INDIVIDUALS WHO ARE MORE PSYCHE - SUSCEPTIBLE THAN THE OTHERS. SOME INDIVIDUALS ARE PRONE FOR PARTICLE INFLICTIONS, AND THE OTHERS ARE NOT.

A FEW FACTS ABOUT PARTICLE CLOUD TRANSMISSION:

1- TRANSMISSION OF NORMAL S. Y. - F.P. – I.I. – P. cl. FROM ONE TO OTHER, PRODUCE NORMAL PSYCHE, OR NORMAL INDIVIDUAL.

2 - TRANSMISSION OF ABNORMAL S.Y. –F.P. –I.I. –P. cl. PRODUCE ABNORMAL PSYCHE.

3 - TALKING IS EXCHANGING PARTICLE CLOUDS (S. Y.- F.P. – I.I. – P. cl.) BETWEEN INDIVIDUALS THROUGH EXOGENOUS PARTICLE CLOUD CIRCULATION SYSTEMS, THE PARTICLE CLOUDS CIRCULATE BETWEEN INDIVIDUALS IN PARTICLE CLOUD CIRCULATION SYSTEMS CIRCUITS.

4- TRANSMISSION OF NORMAL PARTICLE CLOUDS FROM NORMAL PARTICLE CLOUD DONERS TO NORMAL RECIPIENTS. PRODUCE NORMAL PSYCHE.

5 - TRANSMISSION OF ABNORMAL PARTICLE CLOUDS FROM SICK DONERS TO NORMAL RECIPIENTS, PRODUCE SICK PSYCHE MOSTLY.

6 - ABNORMAL S. Y. T. E. – F.P. – I.I. – P. cl. TRANSMISSION IS CAUSE OF MENTAL DISORDERS

7 - PSYCHIARTRIC DISEASES ARE ABNORMAL PARTICLE CLOUD TRANSMITTED DISORDERS.

8- NORMAL PARTICLE CLOUD COMBINATION WITH INTERNAL ELECTRONS NUCLEONS PARTICLE COMPOUNDS PRODUCE NORMAL CONSTRUCTION CNS ELECTRONS NUCLEONS. AND NORMAL S.Y. – F.P. – I.I. – P.cl. WILL CIRCULATE INSIDE CNS ELECTRONS NUCLEONS PARTICLE CIRCULATION SYSTEMS CENTER.

9- ABNORMAL PARTICLE CLOUDS COMBINATIONS WITH CNS ELECTRON NUCLEONS PARTICLE COMPOUNDS CONSTRUCT ABNORMAL CNS PARTICLE CLOUD COMPOUND CONSTRUCTIONS ELECTRON NUCLEON. MANIFEST ITSELF AS ABNORMAL SYMPTOMS AND SIGNS.

CAUSE OF PSYCHIATRIC DISOREDERS ARE PROLONG TRANSMISSION OF ABNORMAL S. Y. – F.P. – I.I. – P. cl. FROM ABNORMAL PARTICLE CLOUD INFLICTED SICK PSYCHIATRIC PATIENTS, TO NORMAL CHILDREN, STUDENTS, INDIVIDUALS, THE PARTICLE-CLOUD DONER AND TRANSMITTER INDIVIDUALS CAN BE ANY ONE SUCH AS: PARENTS, TEACHERS, INSTRUCTERS, RULERS, SOCIO – POLITICAL OR RELIGIOUS INDIVIDUALS, ETC.

THE ABNORMAL S. Y. – F.P. – I.I. – P. cl. AIRBORN TRANSMIT FROM SICK TO NORMAL INDIVIDUALS RECIPIENT'S SENSARY ORGANS, ELECTRONS, NUCLEONS ATTRACT PARTICLE CLOUDS, AND SENSARY ORGANS THROUGH NANO-FIBERS OF NEURAL CONSTRUCTIONS CARRY INCOMING ABNORMAL PARTICLE CLOUDS INTO RECIPIENTS CNS SENSARY CENTERS ELECTRONS NUCLEONS, AND CONSTRUCT THE PARTICLE CLOUD COMPOUND CONSTRICTIONS IN DIFFERENT CNS CENTERS ELECTRONS NUCLEONS CONSTRUCTIONS.

INCOMING PARTICLE CLOUDS INTER INTO CNS SENSARY INTERNAL ELECTRONS, NUCLEON SUBSYSTEM-UNITS CHEMICAL LAB.S, AND THE EXOGENOUS PARTICLE CLOUDS COMBINE WITH INTER ELECTRON NUCLEON PARTICLE COMPOUNDS, AND PRODUCE S. Y. – F.P. – I.I. – P. cl. COMPOUNDS

AND CONSTRUCT RECIPIENTS ELECTRONS NUCLEONS PARTICLE COMPOUNDS CONSTRUCTIONS, WITH INCOMING ABONORMAL PARTICLE CLOUD COMPOUNDS, ABNORMAL SONIC PARTICLE CLOUDS CONSTRUCT ABNORMAL SONIC PARTICLE COMPOUNDS (S. F.P. – I.I. – P. cl. _ COMPOUNDS) IN AUDITARY CNS CENTERS ELECTRONS NUCLEONS ABNORMALLY, AND ABNORMAL LIGHT PARTICLE CLOUDS CONSTRUCT ABNORMAL LIGHT PARTICLE COMPOUNDS IN VISION CENTERS ELECTRONS NUCLEONS AND CONSTRUCT THE VISION CENTERS INTER ELECTRON-NUCLEON PARTICLE COMPOUND CONSTRUCTIONS WITH ABNORMAL Y. F.P. – I.I. – P. cl. COMPOUNDS,

THE INTERACTIONS OF THESE ABNORMAL S. Y. – F.P. – I.I. – P. cl. WITH EACH OTHER, BETWEEN DIFFERENT CNS CENTERS, AS WELL AS THEIR INTERACTIONS WITH INCOMING OTHER S. Y. – F.P. – I.I. – P. cl. PRODUCE ABNORMAL PARTICLE CLOUD CURRENTS, THE ABNORMAL PARTICLE CLOUD INTERACTIONS TAKE PLACE THROUGH A. S. I. F.P. –Mol. C.I.C. AND PRODUCE ABNORMAL THOUGHT CURRENTS, AND ABNORMAL PSYCHE, THIS IS PHENOMENON OF ABNORMAL PSYCHE- GENESIS AND ABNORMAL THOUGHT CURRENT – GENESIS PHENOMENONS,

PSYCHE- GENESIS

DIFFERENT KIND PARTICLE CLOUD (S. Y. –F.P. – I.I. – P. cl.) TRANSMISSIONS, PRODUCE DIFFERENT KIND PSYCHE, AND THOUGHT CURRENTS

RMAL PARTICLE CLOUD DONNERS, CONSTRUCT CNS INTERNAL ELECTRON-NUCLEON PARTICLE COMPOUND CONSTRUCTIONS OF PARTICLE CLOUD RECIPIENT, CHILDREN, STUDENTS, INDIVIDUALS INTERNAL ELECTRON-NUCLEON PARTICLE COMPOUND CONSTRUCTIONS, WITH ABNORMAL S. Y. – F.P. –I.I. – P. cl. – COMPOUND CONSTRUCTIONS.

THE PSYCHOTIC DELUSIONAL PARTICLE CLOUD (S.Y. – F.P.- I.I. – P. cl.) TRANSMISSION, TO RECIPIENTS PRODUCE PSYCHOSIS TYPES PARTICLE CLOUD COMPOUND CONSTRUCTIONS, INSIDE RECIPIENTS CNS ELECTRONS NUCLEONS. THE DEPRESSIVE PARTICLE CLOUD (S.Y. – F.P. – I.I. – P. cl.) TRANSMISSIONS PRODUCE DEPRESSION TYPE PARTICLE CLOUD INFLICTIONS ATOMS. THE NEUROTICS NERVOUS PARTICLE CLOUD TRANSMITTERS TRANSMIT NEUROSIS TYPES PARTICLE CLOUDS TO RECIPIENTS AND CONSTRUCT NEUROTIC KIND PARTICLE CLOUD COMPOUNDS INSIDE THE RECIPIENTS CNS INTERNAL ELECTRONS NUCLEONS PARTICLE COMPOUNDS. AND SO ON,

THE NORMAL PSYCHE PARENTS, TEACHERS, INSTRUCTORS, TEACH AND TRANSMIT NORMAL PARTICLE CLOUDS (S. Y. – F.P. – I.I. – P. cl.) TO RECIPIENT CHILDREN, STUDENTS, INDIVIDUALS, AT HOME, SCHOOL, UNIVERSITY, AND PUBLIC PLACES, AND PRODUCE NORMAL CNS S. Y. – F.P. –I.I. – P. cl. -PARTICLE COMPOUND COMSTRUCTIONS IN RECIPIENT CHILDRENS, STUDENTS, AND INDIVIDUALS,

THE DEVIATED AND CRIMINAL PARENTS, TEACHERS, SCHOOLS, POLITITIONS, UNIVERSITIES, INSTRUCTORS,

TRANSMIT DEVIATED AND CRIMINAL TYPES OF PARTICLE CLOUDS (S. Y. – F.P. – I.I. – P. cl.) TO RECIPIENT STUDENTS, CHILDREN, PARTIES, SOCIO-POLITICAL SOCIETIES, AND INFLICT NORMAL INDIVIDUALS WITH ABNORMAL DEVIATED OR CRIMINAL PARTICLE CLOUDS (S. Y. – F.P. – I.I. – P. cl.) AND CONSTRUCT RECIPIENT NORMAL OR ABNORMAL INDIVIDUALS INTER CNS INTER ELECTRON, NUCLEON, PARTICLE COMPOUND CONSTRUCTIONS, WITH ABNORMAL DEVIATED TYPES S.Y. – F.P. – I.I. – P. cl. –COMPOUND CONSTRUCTIONS AND PRODUCE CRIMINALS AND DEVIATES,

SOME INDIVIDUALS POSSESSES REMARKABLE PSYCHOGENESIS POWERS AND CAPABILITIES THAN OTHER, SOME ABNORMAL PARTICLE CLOUD INFLICTED INDIVIDUALS, WILL TRANSMIT MENTAL DISEASE TO OTHERS INDIVIDUALS, EVEN IN SHORT CONTACT, IN CONTRAST IN MOST OTHERS TRANSMISSION IS VERY SLOW PROCESS, PSYCHE-GENESIS CAPABILITIES OF DIFFERENT INDIVIDUALS ARE DIFFERENT FROM EACH OTHER.

MOST INDIVIDUALS TRANSMIT THE KIND OF S.Y – F.P. – I.I. –P. cl. FROM SELF, WHICH THEIR INTERNAL CNS INTER ELECTRONS –NUCLEONS PARTICLE CLOUD COMPOUNDS CONSTRUCTIONS, HAVE BEEN MADE AND CONSTRUCTED FROM THOSE KIND S.Y. –F.P. – I.I. – P. cl. COMPOUNDS.

THE INDIVIDUALS WHOSE INTERNAL CNS INTER ELECTRON-NUCLEON PARTICLE COMPOUND CONSTRUCTIONS, HAS BEEN PRODUCED BY USE OF PSYCHOTIC S. Y. – F.P. – I.I. – P. cl. COMPOUNDS CONSTRUCTIONS, THESE PATIENTS ARE

PSYCHOSIS TRANSMITTERS, THE PSYCHOSIS TRANSMITTERS MOSTLY PRODUCE AND TRANSMIT, AIRBORN ABNORMAL PSYCOSIS PARTICLE CLOUDS (S.Y. – F.P. – I.I. P. cl.) AND CAUSE PRODUCTIONS OF PSYCHOSIS IN RECIPIENTS, CHILDRENS, AND INDIVIDUALS.

IN CONTRAST THE PARENTS, TEACHERS, INSTRUCTORS, WHOSE CNS INTER ELECTRON – NUCLEON PARTICLE COMPOUND CONSTRUCTIONS, THAT HAVE BEEN CONSTRUCTED FROM DEPRESSIVE, OR ANXIETY NERVOUS TYPES, OR OTHER KIND PARTICLE CLOUDS (S.Y. – F.P. – I.I. P. cl.) COMPOUNDS CONSTRUCTIONS, THESE INDIVIDUALS TRANSMIT FROM SELF DEPRESSION, NEUROSIS, AND ANXIETY KIND PARTICLE CLOUDS TO OTHERS, AND IN RECIPIENTS UNDER A. S. I. F.P. Mol. C.I.C. INSIDE THEIR CNS ELECTRONS NUCLEONS PARTICLE CLOUDS COMBINE WITH INTERNAL ELECTRON NUCLEON PARTICLE COMPOUNDS, AND PRODUCE DEPRESSIVE NEW (S.Y. – F.P. – I.I. – P. cl.) PARTICLE CLOUD – COMPOUNDS, OR NEUROTIC S.Y. –F.P. – I.I.—P.cl. COMPOUNDS AT INTER ELECTRON-NUCLEON CONSTRUCTIONS, AND CREATE IN RECIPIENTS THE SAME KIND PARTICLE CLOUD TRANSMITTED DISEASES. AND THESE ARE NEW PARTICLE CLOUD INFECTED PATIENTS.

PARTICLES, PARTICLE- CLOUDS, DIFFERENT TYPES NORMAL OR ABNORMAL TYPES FUNDAMENTAL PARTICLE INFORMATION AND IMAGE PARTICLE CLOUDS (F.P. – I.I.- P. cl.) TRAVEL AND TRANSIT AIRBORN, INFLICT RECIPIENT LIVING THINGS, WITH SAME KIND NORMAL OR ABNORMAL PARTICLE CLOUD TRANSMITTED (S. Y. – F.P. – I.I. – P. cl.) TRANSMITTED

INFECTIONS, PRESENTLY KNOWN AS MENTAL CONDITIONS.

THE MENTAL DISORDERS ARE PARTICLE TRANSMITTED DISEASES, AND PARTICLE INFECTED PATIENTS.

ANIMAL GENESIS

CONSTRUCTIONS OF BIOLOGICALLY NEEDED BODY SYSTEMS AND ORGANS FOR LIVING THINGS

UNDER MOLECULAR EVOLUTION ORDERS
AND THROUGH A.S.I. Mol. C.I.C.

DURING THE PARTICLE MOLECULAR EVOLUTION, THE PARTICLE INTELLIGENCE SYSTEMS INVENTED AND CONSTRUCTED DIFFERENT BIOLOGICAL DEVICES, CONSTRUCTED DIFFERENT BODY ORGANS, AND BODY SYSTEMS, PARTICLES FOR PERFORMING ENORMOUS DIFFERENT KINDS EXISTING LIVING THINGS DIFFERENT BIOLOGICAL FUNCTIONS, WHICH THESE TASKS ARE NECESSARY AND NEEDED FOR OPERATION OF A BIOLOGICAL MACHINES.

THE PARTICLES CONSTRUCTED NEEDED BODY SYSTEMS SUCH AS: FLUIDS -FOODS PUMPERS CIRCULATIONS SYSTEMS BIO-DEVICES, BONY – MUSCULAR MOVEMENT BIO- DEVICES, DEFLATING –INFLATING GAS EXCHANGE BIO-DEVICES, AND MANY OTHER NEEDED BODY SYSTEMS AND ORGANS WHICH EACH ACHIEVE ONE OR MANY DIFFERENT TASKS, ALL OF BODY SYSTEMS AND ORGANS CONSTRUCTED

THROUGH PARTICLE INTELLIGENCE SYSTEMS, AND ATOMS, BIOMOLECULES, CELL'S INTELLIGENCE SYSTEMS COOPERATING WITH EACH OTHER, UNDER THE MOLECULAR EVOLUTION ORDERS, AND A.S. I. F.P. Mol. C.I.C.

FOR COMMUNICATIONS OF DIFFERENT INDIVIDUALS CNS ELECTRONS NUCLEONS WITH EACH OTHER, THE PARTICLES CONSTRUCTED BIO-DEVICES: SUCH AS AIRBORN EXOGENOUS PARTICLE CLOUD CIRCULATION SYSTEMS (EX. -PCS), WHICH EXOGENOUS PARTICLE CIRCULATION SYSTEMS ENABLE DIFFERENT LIVING THING SPECIES EXCHANGE KNOWLEDGES BETWEEN EACH OTHER, ALSO PARTICLES CONSTRUCTED INDIGENOUS PARTICLE CIRCULATION SYSTEMS (PRESENTLY KNOWN AS NERVOUS SYSTEMS). THE ELECTRONS NUCLEONS OF DIFFERENT LIVING THINGS SPECIES THROUGH THESE PCS – BIO-DEVICES COMMUNICATE WITH EACH OTHER.

NANO- COMMUNICATION BETWEEN TOTAL BODY ELECTRONS –NUCLEONS POPULATIONS

PARTICLE CLOUS (S. Y. E. T. – F.P. – I.I. – P. cl.) CIRCULATION SYSTEMS

THERE ARE TWO CLASSES PARTICLE CLOUD CIRCULATING SYSTEMS:

1- INTERNAL BODY PARTICLE CLOUD CIRCULATION SYSTEMS OR INDIGENOUS PARTICLE CLOUD CIRCULATION SYSTEMS: THIS SYSTEM PARTICLE CLOUD CIRCULATION

SYSTEM FLOWS, AND INTER-CONNECTS THE TOTAL BODY ELECTRON-NUCLEON POPULATIONS TO EACH OTHER, IT IS NANO-CIRCULATION SYSTEMS, IT'S MICRO AND MACRO CONSTRUCTIONS FORMS ARE PRESENTLY KNOWN AS NERVOUS SYSTEM TODAY.

2 – THE EXOGENOUS PARTICLE CLOUD CIRCULATION SYSTEMS: THIS IS THE PARTICLE CLOUD (S. Y. E. T. –F.P. – I.I. –P. cl.) CURRENTS CIRCULATION SYSTEMS THAT FLOW AIRBORN BACK AND FORTH BETWEEN CNS ELECTRON NUCLEONS OF DIFFERENT COMMUNICATING INDIVIDUALS, WHICH BY ORDINARY PEOPLE ARE KNOWN AS TALKING AND LISTENING PROCEDURES. THIS EXOGENOUS PARTICLE CLOUD CIRCULATION SYSTEM ALSO EXIST BETWEEN THE OTHER LIVING THINGS DIFFERENT SPECIES AS WELL. THERE ARE THIRD OTHER TYPES PARTICLE CIRCULATION SYSTEMS BETWEEN NON-LIVE NANO-UNITS ALSO.

EXOGENOUS PARTICLE CLOUD CIRCULATION SYSTEMS AND PHENOMENON OF PARTICLE CLOUD EXCHANGE SYSTEMS BETWEEN DIFFERENT CREATUTES

ANIMALS CNS ELECTRON-NUCLEON THROUGH BILLIONS TO TRILLIONS CONCURRENT DIFFERENT A. S. I. F.P. Mol. C.I.C. WHICH TAKES PLACE BETWEEN DIFFERENT CNS CENTERS ELECTRONS NUCLEONS WITH EXPONENTIAL SPEEDS, PRODUCE AND EMITS AIRBORN PARTICLE CLOUDS TO THE OUTSIDE WORLD INTO AIR, THESE PARTICLE CLOUDS TRANSIT IN ALL AIR DIRECTIONS INTO SPACE.

THE RECIPIENTS CNS ELECTRONS NUCLEONS CAPTURE THESE AIRBORN PARTICLE CLOUDS FROM AIR THROUGH EXISTING MULTI DIFFERENT ORGANS AND SYSTEMS, THESE INCOMING S.Y. – F.P. – I.I. – P.cl. INSIDE RECIPIENTS CNS ELECTRONS NUCLEONS ANALIZED, THROUGH TRILLIONS OF DIFFERENT A. S. I. F.P. Mol. C.I.C. WHICH TAKE PLACE WITH EXPONENTIAL SPEEDS OF INTERACTIONS, BETWEEN DIFFERENT CNS CENTERS ELECTRONS AND NUCLEONS, PRODUCE RESPONSE PARTICLE – CLOUDS AND EMIT THOSE RESPONSE - PARTICLE - CLOUDS (S.Y – F.P. – I.I. – P. cl.) INTO OUTSIDE WORLD AS AIRBORN PARTICLE CLOUDS AGAIN. THESE PARICLE CLOUDS CAPTURED BY OTHERS CNS ELECTRONS-NUCLEONS, THESE BACK AND FORTH PARTICLE CLOUD CURRENTS IN CLOSED PARTICLE CLOUD CIRCULATION SYSTEMS CIRCUITS. CIRCULATE BETWEEN DIFFERENT INDIVIDUALS, ANIMALS AS WELL AS NON LIVE ELECTRONS NUCLEONS. WHICH THESE PHENOMENONS ARE EXOGENOUS PARTICLE CLOUD CIRCULATIONS, AND PARTICLE CLOUD EXCHANGE SYSTEMS, AND PRESENTLY PEOPLE CALL IT SPEAKING AND LISTENING.

ABOVE TYPES FUNDAMENTAL PARTICLE CLOUD CIRCULATING SYSTEM ARE EXTERNAL AIRBORN S. Y. – F.P.- I.I. – P. cl. EXCHANGES, BETWEEN TWO OR MORE DONER AND RECIPIENT LIVING THINGS, SUCH AS HUMAN, OR OTHER ANIMAL SPECIES, ALSO THIS PHENOMENON OCCURS BETWEEN NON- LIVE ELECTRONS NUCLEONS PARTICLE TRANSMISSIONS THROUGH AIRBORN PARTICLE CIRCULATIONS SYSTEMS,

3 - THE THIRD SEGMENTS OF EXOGENOUS PARTICLE CIRCULATION SYSTEMS, AND INDIGENOUS PARTICLE CLOUD CIRCULATING SYSTEMS, ARE CONNECTIONS BETWEEN THESE TWO SYSTEMS, WHICH TAKE PLACE BETWEEN TWO ABOVE EXPLAINED PARTICLE CLOUD CIRCULATION SYSTEMS IN JUCTURES, WHEN BOTH SYSTEMS PARTICLE CLOUD CURRENTS ARE CONNECTING BY SYNAPSUS AND JOINING TO EACH OTHER, AND PRODUCING A SINGLE NON INTERRUPTED PARTICLE CLOUD CURRENT, AS A SINGLE CONNECTED CLOSED CIRCUITS.

PSYCHE GENESIS AND THOUGHT CURRENT GENESIS PHENOMENON STORING VIRTUAL PARTICLE CLOUD COPIES OF OUTSIDE WORLD INSIDE CNS ATOMS:

(S. Y. –F.P.- I.I.- P.cl. ARE VIRTUAL PARTICLE CLOUD COPIES OF OUTSIDE WORLD'S -THINGS)

THE CNS INTERNAL ELECTRON NUCLEON INTERACTIONS ABOUT STORED S. Y.-F.P.- I.I.- P.cl. OR VIRTUAL CLOUDS WHICH ALL ARE INFORMATIONS AND IMAGES ABOUT OUTSIDE WORLD SUBJECTS, THESE INTERACTIONS BETWEEN CNS ELECTRONS NUCLEONS MIMICS MOVIES LIKE FUNCTIONS, FROM ACTING, SPEAKING, MOVING, TEACHING, FROM OUTSIDE SUBJECTS, BIOLOGICALLY SENSE AS PSYCHE, THOUGHT CURRENTS.

SONIC AND LIGHT FUNDAMENTAL PARTICLES CONSTRUCT INFORMATION IMAGE PARTICLE CLOUDS COPIES FROM EXISTING THINGS OF EARTH (S.Y. – F.P. – I.I. –P. cl.), SONIC PARTICLES AND LIGHT PARTICLES BOTH ARE SENSIBLE, AND WE CAN SEE LIGHT PARTICLES AND WE CAN HEAR SONIC PARTICLES OF EARTH. THE PRODUCED SONIC PARTICLE AND

LIGHT PARTICLE INFORMATION IMAGE PARTICLE CLOUDS (S. Y. – F.P. – I.I.- P.cl.), WHEN THEY CONSTRUCT THE S. Y. – F.P. – I.I. – P. cl. COMPOUNDS INSIDE CNS ELECTRONS AND NUCLEONS, WHEN THEIR S.Y. – F.P. – I.I. – P.cl CURRENTS INTERACTIONS UNDER A. S. I. F.P. Mol. C.I.C., INSIDE THE BRAIN ELECTRONS NUCLEONS, BIOLOGICALLY SENSED AND PRODUCE THOUGHT CURRENT SYSTEMS AND PSYCHE, THIS IS PHENOMENON OF PSYCHE-GENESIS AND THOUGHT CURRENT –GENESIS.

SONIC – LIGHT FUNDAMENTAL PARTICLE INFORMATION IMAGE PARTICLE –CLOUDS (S.Y.-F.P.- I.I.-P.cl.) CONSTRUCT S. Y. – F.P. – I.I. – PARTICLE CLOUD COPIES FROM EXISTING THINGS OF EARTH, THE EARTH'S LIGHT PARTICLES AND SONIC PARTICLES ARE SENSIBLE.

THE S. Y. – F.P. – I.I. – P. cl. –COMPOUNDS WHICH ARE COPIES OF ENVIRONMENTAL SUBJECTS AND EVENTS INSIDE CNS INTER-ELECTRON NUCLEON IN PARTICLE COMPOUND CONSTRUCTIONS FORMS, AND CONSTRUCT CNS ELECTRON NUCLEONS, THEIR INTERACTIONS IN BRAIN CAN BE SENSED AS EXISTING ORIGINAL SUBJECTS AND EVENTS PLAYING THROUGH THOUGHT CURRENTS AND PSYCHE FORMS. THESE INFORMATION IMAGE PARTICLE CLOUD COPIES INSIDE CNS ELECTRONS NUCLEONS AND THEIR INTERACTIONS WITH EACH OTHER UNDER A. S. I. F.P. Mol. C.I.C., BIOLOGICALLY SENSE AS THOUGHT CURRENTS AND PSYCHE.

COMBINING S.Y. – F.P. – I.I. –P. cl. WITH SILICON'S INTERNAL ELECTRON- NUCLEON PARTICLE COMPOUNDS

STORAGE OF PARTICLE CLOUDS INSIDE SILICON ATOMS.

THIS IS INFORMATION IMAGE PARTICLE CLOUD - STORAGE PHENOMENON, INSIDE SILICON'S NON LIVE ELECTRONS NUCLEONS, IN THE FORMS OF PARTICLE CLOUD COMPOUNDS.

THROUGH COMPUTER TECH. TRANSMIT THE PRODUCED SONIC LIGHT FUNDAMENTAL PARTICLE INFORMATIONS IMAGES PARTICLE CLOUDS (S. Y. –F.P. – I.I. – P. cl.) WHICH ARE PARTICLE CLOUD COPIES FROM ENVIRONMENTAL EXISTING THINGS, INTO SUBSYSTEM UNITS CHEMICAL LAB.S OF SILICON ELECTRONS AND NUCLEONS, UNDER REGENERATIVE A. S. I. F.P. Mol. C.I.C. COMBINE EXOGENOUS INCOMING S. Y. –F.P. – I.I. –P.cl. WITH SILICONS INTER-ELECTRON- NUCLEON PARTICLE COMPOUNDS, AND PRODUCE SONIC – LIGHT FUNDAMENTAL PARTICLE INFORMATION IMAGE PARTICLE CLOUD COMPOUNDS (S. Y. –F.P. – I.I. – P.cl. _ COMP.) THIS PHENOMENON IS STORAGE OF EXOGENOUS S. Y.- F.P. – I.I. – P. cl. INSIDE SILICONS INTER ELECTRONS NUCLEONS, IN THE SONIC – LIGHT FUNDAMENTAL PARTICLE INFORMATION IMAGE PARTICLE CLOUD COMPOUND MOLECULAR FORMS.

THE S. Y. – F.P. – I.I. – P. cl. RETRIEVAL FROM SILICON ELECTRON – NUCLEONS

PHENOMENON OF RETRIEVAL OF RECORDED MEMORIES

REVERSE OF ABOVE INTERACTIONS, UNDER DEGENERATIVE A. S. I. F.P. Mol. C.I.C., CAUSE BREAK DOWN OF LARGE S.Y.- F.P. – I.I.- P.cl. _COMPOUNDS, INTO SMALL MOLECULAR CONSTRUCTING MOLECULES, CAUSE RELEASES OF INFORMATIONS IMAGE PARTICLE CLOUDS TO OUT OF SILICON'S ELECTRONS NUCLEONS INTO AIR, AND AFTER RELEASE OR RETRIEVAL, THE S. Y. –F.P. – I.I. – P.cl. CAN BE PRESENTED ON TV SCREENS. AND THIS IS PHENOMENON OF RETRIEVAL AND RELEASE OF PARTICLE CLOUDS INTO OUTSIDE SPACE.

THE INTELLIGENT PLANTS, AND NON – INTELLIGENT PLANTS PLANTS CENTRAL INTELLIGENCE SYSTEMS CENTERS (C. I. S.) PLANTS PERIPHERAL INTELLIGENCE SYSTEM CENTERS (P. I. S.)

THERE ARE TOO MANY PLANT SPECIES, THAT DO NOT HAVE PLANT INTELLIGENCE SYSTEM CENTERS, THERE ARE SOME PLANTS ONLY POSSESSES PARTIAL AND INCOMPLETE CENTRAL INTELLIGENCE SYSTEM CENTERS, THERE ARE ALSO LARGE NUMBERS OF PLANT SPECIES THAT POSSESSES BOTH THE CENTRAL INTELLIGENCE SYSTEM CENTERS (C. I. S.), AND PERIPHERAL INTELLIGENCE SYSTEM CENTERS (P. I. S.) BOTH.

THERE ARE TWO MAIN INTELLIGENCE SYSTEM CENTERS IN PLANTS:

1 – THE PLANT'S PRIMORDIAL CENTRAL INTELLIGENCE SYSTEM CENTERS (C. I. S.) OR PLANT'S MAIN INTELLECTUAL COMMAND CENTERS SYSTEMS.

2 – THE PLANT'S PERIPHERAL INTELLIGENCE SYSTEM CENTERS (P. I. S.):

THE PERIPHERAL INTELLIGENCE SYSTEM CENTERS ALSO DEVIDES INTO TWO DIFFERENT INTELLIGENCE SYSTEMS:

1 – THE PLANTS SENSARY INTELLIGENCE SYSTEM CENTERS (S. I. S.).

2 – THE PLANTS MOTOR INTELLIGNCE SYSTEM CENTERS (M. I. S.).

> THE SENSARY INTELLIGENCE SYSTEMS, AND
> MOTOR INTELLIGENCE SYSTEM CENTERS

THE CENTRAL INTELLIGENCE SYSTEM CENTER OF PLANTS

THE BIOFRIENDLY VISIBLE OR NON-VISIBLE LIGHT FUNDAMENTAL PARTICLES HAVE VITAL IMPORTANCE FOR PLANTS, THE LIGHT FUNDAMENTAL PARTICLES DAILY SHINE ON FLOWERS, LEAVES, STEMS, FRUITS, DIRECT INTER INTO INTERNAL ELECTRON NUCLEON SUBSYSTEM –UNITS CHEMICAL LAB.S, COMBINE WITH INTER ELECTRON NUCLEON PARTICLE COMPOUNDS, CONSTRUCT PLANTS ELECTRONS AND NUCLEONS LIGHT PARTICLE COMPOUND CONSTRUCTIONS, UNDER THE REGENERATIVE A. S. I. F.P. Mol. C.I.C. (NANO-UNIT GENESIS PROCESS).

THE PLANT'S SENSARY SYSTEM CENTERS, CONTINUOUSLY MONITOR AND REPORT EXISTANCES OF ENVIRONMENTAL SUN SHINE, OR EXISTING ENVIRONMENTAL HAZARDS, SUCH AS BURNING SUN IN MID SUMMERS WHICH MAY CAUSE DEHYDRATION AND SUN BURN, OR EXISTANCES OF FREEZING

COLD IN LATE WINTER, OR IN EARLLY SPRING WHICH THE FLOWERS OPEN MAY DIE, THROUGH FROST BITES, THESE ENVIRONMENTAL INFORMATIONS THROUGH PARTICLE CLOUDS TRANSIT, TO THE HIGHER CENTRAL INTELLIGENCE SYSTENS CENTERS (C.I.S.) FOR THEIR DECISIONS, AND ISSUING THE PROPER TASK – ORDERS, FOR MOTOR INTELLIGENCE SYSTEM CENTERS EXECUTIONS.

FOR EXAMPLE, THE CIS IN DAILY SUN SHINE MAY ORDER THE PLANTS FLOWERS, LEAVE STEMS TURN OR TILT TOWARD THE INCOMING SUN LIGHT DIRECTIONS, OR IN THE CASE OF EXISTING FREEZING COLD, MAY ORDER THE CUSPIDS TO CLOSE AND PROTECT THE PETALS FROM FROST BITS, OR IN BURNING HOT TEMPERATURE THE CIS MAY ORDER THE FLOWERS TO TURN DOWN OR AWAY FROM THE INCOMING SUN LIGHT DIRECTIONS TO REDUCE DEHYDRATIONS AND BURNS, ETC.

THE PLANTS C. I. S. CENTERS, THROUGH THE PLANTS MOTOR INTELLIGENCE SYSTEM CENTERS, MAY C. I. S. ORDERS TO MOTOR INTELLIGENCE SYSTEM CENTER THAT, THE FLOWERS, THE LEAVES, AND STEMS MUST ROTATE AND TURN TOWARD THE DIRECTION OF INCOMING LIGHT FUNDAMENTAL PARTICLES DAILY WHEN THERE IS SUN SHINE. IN ORDER TO CAPTURE THE MAXIMUM AMOUNT NEEDED LIGHT FUNDAMENTAL PARTICLE FROM AIR, AND DELIVER INTO ELECTRONS NUCLEONS CHEMICAL LAB.S, TO BE USED IN A.S.I.F.P. Mol. C.I.C. FOR RE-GENESIS OF NEEDED PLANTS NANO-UNIT STRUCTURES. THE MAIN FUNCTION OF SENSARY INTELLIGENCE SYSTEM CENTERS IS TO REPORT EXISTING

SUN SHINE, OR EXISTING ENVIRONMENTAL HAZARDS SUCH AS FREEZING COLD, OR BURNING HOT SUMMER, ETC. TO THE HIGHER COMMAND INTELLIGENCE SYSTEM CENTERS.

THE ESSENTIAL FUNCTIONS OF PLANTS COMMAND SYSTEMS CENTERS IS TO OPERATE PROTECTIVE AND DEFENSIVE MEASURES, PROVIDE AND MAINTAIN THE CHEMICAL, PHYSICAL, AND BIOLOGICAL NEEDED TASKS, IN NORMAL LEVELS IN THE PLANTS. HELP PLANTS CAPTURE NEEDED LIGHT FUNDAMENTAL PARTICLES, ORDER AND ACHIEVE NORMAL PLANTS FUNCTIONS, AND A. S. I. F.P. Mol. C.I.C., ETC., THE PLANTS C.I.S. -CENTERS IN DAILY BASIS, ORDER THE PLANTS LEAVES, FLOWERS, STEMS, TO TURN TOWARD INCOMING LIGHT PARTICLE DIRECTIONS, TILT, OR ROTATE, AND TURN TOWARD LIGHT PARTICLE DIRECTIONS WHEN THERE IS SUNSHINE. IN ORDER TO COLLECT LIGHT PARTICLES FROM INCOMING DIRECTIONS OF LIGHT ORIGIN PARTICLES, THROUGH THE FLOWERS AND LEAVES AND STEM ROTATIONS OR TILTS, UNDER THE CIS INTELLECTUAL COMMAND SYSTEMS CENTERS ORDERS THE LEAVES, FLOWERS, STEMS TURN TOWARD LIGHT PARTICLE DIRECTIONS, AND ATTRACT THE LIGHT PARTICLES.

IN DAILY BASIS, THE AIRBORN INCOMING LIGHT PARTICLES DIRECTLY INTER INTO INSIDE PLANTS INTER ELECTRON NUCLEON SUBSYSTEM –UNITS CHEMICAL LAB.S, AND UNDER A. S. I. F.P. Mol. C.I.C. COMBINE WITH INTER ELECTRON-NUCLEON PARTICLE COMPOUNDS AND CONSTRUCT LIGHT PARTICLE COMPOUNDS OF PLANTS ELECTRON NUCLEON CONSTRUCTIONS, (PHENOMENONS OF

PARTICLE COMPOUND NEO- GENESIS AND PHENOMENON OF NEW ELECTRON- NUCLEON –GENESIS) ALL OF THESE NANO-TASKS, SENSED AND REPORTED BY S.I.S, AND NECESSARY ORDERS ISSUED THROUGH THE C.I.S., AND THE TASKS EXECUTED BY M.I.S CENTERS. ALL PLANTS DEFENSIVE AND PROTECTIVE MEASURES, DAILY PHYSICAL CHEMICAL BIOLOGICAL FUNCTIONS ALL SENSED BY S.I.S., ORDERS ISSUED BY C.I.S. CENTERS, AND ISSUED ORDERS EXECUTED BY M.I.S. CENTERS OF THE PLANTS. UNDER DIRECT C.I.S. CENTERS ORDERS THE FLOWERS, STEMS, LEAVES, TURN INTO DIRECTION OF SUN LIGHT, THE PLANTS DEFENSIVE AND PREVENTIVE ORDERS ALSO DONE BY CENTRAL INTELLIGENCE SYSTEM CENTERS AND ISSUED ORDERS CARRIED OUT BY M.I.S. CENTERS.

THE S.I.S. CENTERS FUNCTIONS ARE SENSING, AND REPORTING, EXISTING ENVIRONMENTAL CONDITIONS, AS WELL AS REPORTING EXISTANCES OF HARMFUL ENVIRONMENTAL CONDITIONS, SUCH AS DEHYDRATIONS, LACK OF IRRIGATIONS ALL GET REPORT TO C. I. S. CENTERS FOR THEIR DECISIONS AND ORDERS. AND CIS EXECUTE THE ORDERS THROUGH M.I.S. CENTERS.

THROUGH THE PROPER OPERATIONS OF S. I. S., C. I. S. AND M. I. S. CENTERS CLOSELY, THE CARNIVOROUS PLANTS CAN DO INSECT HUNTING THROUGH QUICK INTELLIGENCE SYSTEMS OPERATIONS. SENSARY SYSTEMS CAN PINPOINT EXACT INSECT LOCATIONS ON LEAVES, AND C.I.S. CAN ISSUE ORDER AT EXACT TIME, AND M. I. S. CAN EXECUTE THE HUNTING ORDERS ACCORDINGLY. AND CATCH FAST MOVING INSECT

IN EXACT LOCATION AND EXACT MOMENTS, AND DELIVER THE HUNTED LIVE INSECTS FOR PLANTS NUTRITION AS ORDINARY DAILY FOOD.

THE PARTICLE CLOUD CURRENTS, AND PARTICLE CIRCULATION SYSTEMS (PCS) IN PLANTS

THE PRIMARY P. C. S. CONNECTS THE THREE SENSORY - CENTRAL - MOTOR INTELLIGENCE SYSTEM ELECTRONS-NUCLEONS TO EACH OTHER, THROUGH BI-DIRECTIONAL SEPARATE EFFERENT AND AFFERENT PARTICLE CIRCULATION CURRENTS, WHICH THESE TWO DIRECTION CURRENTS, ONE PARTICLE CLOUD CURRENTS RUNS, IN OPPOSITE DIRECTION OF THE OTHER PARTICLE CLOUD CURRENT.

ALSO, THE SECONDARY P. C. S. CONNECTS THE PLANTS: STEMS, FLOWERS, LEAVES, ROOTS, FRUITS TOTAL ELECTRON – NUCLEON POPULATIONS TO EACH OTHER. AND ALL OF THESE SECONDARY P. C. S. CURRENTS FINALLY CONNECTS TO MAIN PRIMARY PLANTS PARTICLE INTELLIGENCE SYSTEMS CENTERS ELECTRONS –NUCLEONS POPULATIONS P. C. S. CURRENTS FROM THE OTHER SIDE,

THROUGH THESE TWO DIFFERENT DIRECTION PRIMARY AND SECONDARY PARTICLE CLOUD CIRCULATIONS SYSTEMS CURRENTS THE ENTIRE PLANTS ELECTRONS NUCLEONS POPULATIONS GET CONNECTION TO EACH OTHER. ABOVE IS P.C.S CONSTRUCTIONS BRIEFLY. WHICH THROUGH THESE PCS THE PLANTS INTELLECTUAL CENTERS ARE ABLE TO MANAGE THE OPERATIONS OF THE ENTIRE PLANTS BODY

ELECTRONS NUCLEONS FUNCTIONS.

CLASSIFICATIONS OF PLANT SPECIES UNDER PLANT'S PARTICLE INTELLIGENCE SYSTEMS:

UNDER THIS CLASSIFICATIONS THE DIFFERENT PLANTS SPECIES DIVIDE INTO THREE DIFFERENT CLASSES: EITHER PLANTS HAVE ALL THREE SENSARY- CENTRAL-MOTOR IN- TELLECTRUAL SYSTEMS CENTERS ALL. OR THE PLANTS SPE- CIES WHO DO NOT HAVE ANY KIND INTELLIGENCE SYSTEM CENTERS AT ALL. ALSO THERE ARE THIRD CLASS LARGE NUMBERS OF PLANTS SPECIES WHO POSSESSES INTER- MEDIATE AND INCOMPLETE PLANT INTELLIGENCE SYSTEM CENTER, WHICH EXPLAING CHARACTORS OF THIS THIRD CLASS IS TOO LARGE, AND DO NOT FIT TO BE EXPLAINED HERE.

THE CLASSIFICATION OF PLANTS UNDER PARTICLE INTELLI- GENCE SYSTEMS:

1 - THE PLANT SPECIES WHO POSSESS C. I. S., AND P.I.S.

2 - THE PLANTS SPECIES THAT DO NOT HAVE C. I. S. AND P.I.S.

3 - THE PLANTS SPECIES WHO POSSESS PARTIAL AND IN- COMPLETE C. I. S, PIS, CIS.

THE PLANT SPECIES WHO POSSESS C.I.S. AND P.I.S.

PLANTS PRIMORDIAL INTELLIGENCE SYSTEMS CENTERS COMPOSED FROM THREE MAJOR DIFFERENT INTELLIGENCE SYSTEM CENTERS AS FOLLOWING,

1- PLANTS CENTRAL INTELLIGENCE SYSTEM CENTERS (C. I. S.), 2 - PLANTS SENSARY INTELLIGENCE SYSTEMS CENTERS (S. I. S.), 3 - PLANTS MOTOR INTELLIGENCE SYSTEMS CENTERS (M. I. S.),

THESE ADVANCED PLANT SPECIES POSSESS CAPABILITIES THROUGH SENSARY INTELLIGENCE SYSTEMS SENSE, CAPTURE AND RECEIVE LIGHT INFORMATION IMAGE PARTICLE CLOUDS FROM AIR IN REGARD TO OUTSIDE WORLD ENVIRONMENTAL CONDITIONS AND INSIDE CENTRAL INTELLIGENCE SYSTEM CENTERS ANALIZE THE SENSARY SYSTEMS REPORTED CONDITIONS AND DECIDE, ORDER, RESPOND ACCORDING TO CONDITIONS AND CARRY ORDERS TO PLANTS MOTOR INTELLECTUAL SYSTM FOR EXECUTION,

THE SENSARY INTELLIGENCE SYSTEMS ELECTRONS NUCLEONS THROUGH THEIR HIGH GRAVITON FORCES, ATTRACT AIRBORN LIGHT PARTICLES AND LIGHT FUNDAMENTAL PARTICLE INFORMATIONS- IMAGES PARTICLE- CLOUDS (Y. F.P. –I.I.- P.cl.) FROM AIR, AND TRANSIT INTO PLANTS CENTRAL INTELLIGENCE SYSTEM CENTERS ELECTRONS –NUCLEONS, THE INCOMING Y. F.P. –I.I.- P.cl. DIRECTLY INTER INTO INTER ELECTRON- NUCLEON SUBSYSTEM UNITS PLANTS CENTRAL INTELLIGENCE SYSTEMS AND COMBINE WITH INTER

ELECTRON-NUCLEON PARTICLE COMPOUNDS UNDER A. S. I. F.P. Mol. C.I.C. AND PRODUCE Y. –F.P.- I.I. – P. cl. _COMPOUNDS, AND THESE PRODUCED PARTICLE CLOUD COMPOUNDS CONSTRUCT THE C. I. S. ELECTRONS NUCLEONS CONSTRUCTIONS, (NEO-ELECTRON-NUCLEON GENESIS).

THE DIFFERENT CIS PRE-EXISTING CONSTRUCTED ELECTRONS –NUCLEONS THROUGH THESE C. I. S. INFORMATIONS AND IMAGES DECIDE IN REGARD TO NEW INCOMING PARTICLE CLOUDS AND DIAGNOSE THE SITUATIONS AND ISSUE THE C.I.S. ORDERS TO M. I. S. ELECTRONS NUCLEONS TO BE EXECUTED ACCORDINGLY, THROUGH THESE COMPLEX A. S. I. F.P. Mol. C.I.C., PARTICLE CLOUD COMPARISIONS AND THE C.I.S. ELECTRONS-NUCLEONS DECISIONS OVER THE S.I.S. REPORTS.

OVER INFORMATIONS OF S. I. S. SENSORY PARTICLE CLOUDS REPORTS, THE C.I.S. DECIDES ITS ORDERS, AND ISSUE ORDERS TO M. I. S., THROUGH COMPARISON OF ENORMOUS PRE-EXISTING STORED Y. –F.P. – I.I. – P. cl. COMPOUNDS, WHICH ALL PREVIOUSLY HAVE BEEN STORED INSIDE THE DIFFERENT ELECTRONS NUCLEONS, SIMILAR AN ENCYCLOPEDIA OF INFORMATIONS, FOR COMPARISONS AND DECISION MAKING BY C. I. S. ELECRONS NUCLEONS.

THE PRODUCED ORDERS THROUGH OUTGOING F.P.- I.I. – P.cl. CURRENTS CIRCULATION SYSTEMS TRAVEL TO PLANTS MOTOR INTELLIGENCE SYSTEM CENTERS (M.I.S.) AND MOTOR INTELLIGENCE CENTERS EXECUTE C.I.S. - ORDERES ACCORDING ISSUED ORDERS THROUGH PLANTS

LEAVES- FLOWERS- STEMS ALL ACT ACCORDING ISSUED ORDERS, ISSUED ORDERS MAKES FLOWERS LEAVES STEMS TURN ROTATE ACCORDING S. I. S. REPORTS AND C. I. S. ORDERS, AND M.I.S. SECODARY ORDERS OVER THE STEMS, LEAVES. FLOWERS, AND ROOTS THAT HOW THEY MUST ACT, ROTATE, TILT, OR NOT DO ANY THING.

PLANTS INTELLIGENCE SYSTEM'S FUNCTIONS

EXAMPLES FROM PHYSIOLOGICAL FUNCTIONS UNDER INTELLIGENCE SYSTEM CENTERS.

1 – IN LATE WINTER, OR EARLY SPRING, WHEN WARM WEATHER TRIGGER FLOWERING PLANTS, IN CASES OF SUDDEN FREEZING EPISODES, THE S. I. S. SENSE EVENTS, AND REPORT FREEZING COLD, POSSIBLE FROST BITE TO C. I. S., UNDER ANALYSIS AND DECISSIONS THE C. I. S. ORDERS TO THE M. I. S., TO CLOSE CUSPIDS, AND PROTECT PETALS, OVARIES, STAMEN, FROM FROST BITES, IN THE PLANT SPECIES WHO POSSESS INTELLIGENCE SYSTEMS CENTERS, AND M.I.S. CARRIES ORDERS OF C.I.S. ACCORDINGLY.

IN MANY PLANT SPECIES WHO DO NOT POSSESS INTELLIGENCE PLANT SYSTEMS CENTERS, THESE SPECIES FLOWERS STAY WIDELY OPEN IN FREEZING COLD, AND SUFFER FROST BITE AND DIE. BECAUSE THEY COULD NOT SENSE, BECAUSE THEY DO NOT POSSESS INTELLIGENCE SYSTEMS, TO SENSE AND CLOSE CUSPIDS, PROTECT PETALS OVARIES STAMEN FROM FROST BITES.

2 - IN NORMAL CLIMATE, WHEN THERE IS SUN SHINE, DIFFERENT LIGHT FUNDAMENTAL PARTICLES ARE PLENTY AROUND AIRBORN COMING FROM PLANET SUN, UNDER THESE CONDITIONS, AS PLANTS NEEDS ALL KIND LIGHT FUNDAMENTAL PARTICLES TO CONSTRUCT THEIR INTERNAL ELECTRON NUCLEON PARTICLE COMPOUND CONSTRUCTIONS, THE S.I.S. SENSE SITUATIONS AND REPORT STATUS TO UPPER INTELLIGENCE SYSTEM CENTERS, THE C.I.S. ISSUE ORDERS THAT ALL FLOWERS, LEAVES, FRUIT'S INTERNAL ELECTRON-NUCLEON PARTICLE COMPOUNDS CONSTRUCTIONS THROUGH CAPTURING NEEDED LIGHT PARTICLES FROM SURROUNDING AIR.

THE SENSARY SYSTEMS (S. I. S.) SENSE LIGHT PARTICLES, TEMPERATURES AND REPORTS PARTICLES DIRECTIONS, THROUGH Y. -F.P.- I.I. – P.cl. CURRENTS WHICH CARRIED TO HIGHER PLANTS CENTERAL INTELLIGENCE SYSTEMS CENTER (C. I. S.), IN RESPONSE TO S. I. S. FINDINGS, THE CENTERAL INTELLIGENCE SYSTEM CENTERS ISSUE PROPER ORDERS TO MOTOR INTELLIGENCE SYSTEMS (M. I. S.), TO EXECUTE ORDERS, TURNING FLOWERS AND LEAVES IN DIRECTION OF LIGHT PARTICLE TOWARD SUN, TO CAPTURE MAXIMUM NEEDED QUATITIES LIGHT PARTICLES FOR ELECTRON-NUCLEON –GENESIS WHICH ARE SURVIVAL FOR PLANT'S LIFE.

3 - IN EXTREME BURNING SUMMER TEMPERATURES, THE SENSARY INTELLIGENCE SYSTEMS SENSE DANGEROUS BURNING, DRYINGS, LETHAL, DEHYDRATING CONDITIONS, AND REPORT TO C. I. S. THE CENTRAL INTELLIGENCE SYSTEM CENTER ORDER TO MOTOR INTELLIGENCE SYSTEMS

CENTERS TO TURN THE FLOWERS FACE DOWN, AND AWAY FROM EXPOSURE TO DIRECT HOT BURNING SHINING LIGHT PARITCLES, AND PREVENT AND AVOID FROM DIRECT BURNING LIGHT PARTICLES EXPOSURES.

IN FEW ANOTHER CASES OR SPECIES WHO POSSESS BETTER DEFENSIVE AND IMMUNITY SYSTEMS UNDER THESE CONDITIONS WHEN THE PLANTS RECEIVE PROPER HELPS SUCH AS IRRIGATIONS, ETC., FLOWERS, LEAVES START THEIR ORDINARY FUNCTIONS TURN TOWARD SUN TO COLLECT MORE LIGHT PARTICLES AND WHICH THEY NEEDS FOR DAILY USE,

IN SPECIES WHO DO NOT POSSESS THE PLANT INTELLIGENCE SYSTEM CENTERS, THESE SPECIES GET HARMED FROM HEAT STROKE, BURNS, DEHYDRATIONS, MUCH EASIER THAN THOSE WHO HAVE PRIMARY INTELLIGENCE SYSTEM CENTERS.

PHENOMENON OF PLANT IMMUNITY SYSTEMS AND DEFENSIVE SYSTEMS

PLANT'S CHILDHOOD ERA, THEREAFTER PLANT'S AGING ERA EFFECTS, AND LIVING IN GERIATRIC AGES, AND GRADUAL SLOWING DOWN OF PLANTS INTELLIGENCE SYSTEMS, DEFENSE SYSTEMS, IMMUNITY SYSTEMS:

THE YOUNG HEALTHY FLOWERS AND PLANTS POSSESS MUCH BETTER DEFENSE SYSTEMS AND IMMUNITY OPERATIONS AND THEIR SENSARY SYSTEMS CENTRAL INTELLIGENNCE SYSTEMS AND MOTOR SYSTEMS ALSO FUNCTION SHARPER AND MUCH BETTER THAN OLDER OR SICK LOOKING PLANTS,

THE ABOVE PHENOMENON IN SOME SPECIES, ALSO ARE BETTER THAN THE OTHER SPECIES WHO DOT POSSESS OR NOT DEVELOPED THESE SYSTEMS. ADDITIONALLY, IN OLDER AGE PLANTS, EVEN THE CENTRAL INTELLIGENCE SYSTEMS OF PLANTS SLOWS DOWN, AND FUNCTIONS CAN NOT BE COMPARED TO YOUNG AGE PLANTS AT ALL.

IN EARLY SPRING WHEN OLD AND YOUNG AGE PLANTS START

FLOWERING, THE PROTECTIVE MOVEMENTS OF FLOWERS AND LEAVES, STEMS AND INTELLIGENCE SYSTEMS FUNCTIONS, IN COMPARING THE OLD PLANTS, WITH YOUNG PLANTS PERFORMANCES ALL ARE DIFFERENT, DETERIORATE AND CHANGES IN VARIABLE DEGREES AT OLDER PLANTS, MOST FUNCTIONS IN OLDER PLANTS BECOME MUCH SLOW AND WEAKER FUNCTIONS THAN, THE YOUNGER PLANT, THE YOUNG PLANTS PERFORM EVERY FUNCTIONS ACCURATE, SHARP, AND FAST, IN OLD PLANTS IT LOOKS CLEARLY OLDER PLANTS EVEN MAY FORGET TO DO THEIR JOBS IN MANY FUNCTIONS AS THEY WERE DOING IN YOUNG AGES.

WHEN COMPARING THE SAME SPECIES YOUNG AND OLD PLANTS NEXT TO EACH OTHER IN A FREEZING OR SNOWING EARLY SPRING MORNING, THE GENERAL PROTECTIVE FUNCTIONS OF YOUNG AGE PLANTS ARE FAR BETTER, THE YOUNG PLANTS START CLOSING CUSPIDS WHEN CLIMATE CHANGES TO FREEZING, BUT THE OLDER PLANTS EVEN HAVE NOT STARTED TO DO THE SAME MAY THEIR INTELLIGENCE SYSTEMS ALSO DETERIORATING, FORGETING AND SLOWING DOWN AND NOT FUNCTIONING AT ALL.

EVEN IN SAME AGE PLANTS ALSO ARE A LOT OF DIFFERENT TYPES VARIABLES, THE IMMUNITY SYSTEMS OF SAME SPECIES PLANTS WITH SAME AGES IN ONE PLANT COMPARING TO OTHER, THE FUNCTIONS ALSO ARE VARIED, ONE DO BETTER THAN THE OTHER.

SOME PLANTS DO NOT POSSESS

INTELLIGENCE SYSTEM CENTER

CONTROL OF STEM CELL'S EXPONENTIAL GROWTH, AND CELL FUNCTIONS, ALL UNDER PARTICLE INTELLIGENCE SYSTEMS:

THERE ARE A LOT OF PLANT SPECIES, WHO DO NOT HAVE INTELLIGENCE SYSTEM CENTERS, AND HAVE NOT DEVELOPED FUNCTIONING PLANT INTELLIGENCE SYSTEMS. THERFORE IN THESE SPECIES, THE DESCRIPTION OF PLANT FUNCTIONS UNDER C. I. S., S.I.S., M.I.S., ARE NOT EXISTED AT ALL. ALSO THERE ARE MANY PLANT SPECIES ONLY HAVE PARTIALLY DEVELOPED INTELLIGENCE SYSTEMS, AND INCOMPLETE INTELLIGENCE SYSTEM CENTERS, THESE SPECIES RESPONSES ARE NOT GOINING TO BE THE SITUATIONS DESCRIBED ABOVE. WHICH WAS ABOUT ADVANCED PLANT SPECIES, WITH FULL DEVELOPED CENTERAL INTELLIGENCE SYSTEM CENTERS.

IN THOSE SPECIES WITH EXISTING C. I. S. IN EARLY SPRING DURING FAST- GROWTH, STEM CELL'S PHASE, WHEN THE NEWLY BORN, EMBRYONIC STEM CELLS START THEIR EXPONENTIAL SPEED MULTIPLICATIONS, AND DUPLICATION PHASE GROWTH, ALL EMBRYONIC PHASES TAKE PLACE UNDER SENSARY REPORTS OF S.I.S. AND ORDERS OF C.I.S., AND THE M.I.S. HAS TO CARRY OUT THE CENTRAL INTELLIGENCE SYSTEMS ORDERS ACCORDINGLY, HOW THE STEM CELLS MUST FUNCTION., AND MUST COORDINATE THEIR FUNCTIONS ACCORDING CELL'S, ATOM'S, INTELLIGENCE SYSTEM'S ORDERS UNDER FUNDAMENTAL PARTICLE OPERATED C. I. S. CENTERS.

AT EARLY STAGE EMBRYONIC PLANT STEM CELL GROWTH PHASES WHEN THE FLOWERS AND BUDS OPENS DURING THE DAY, FOR PROTECTION OF EARLY STAGE PETALS, OVARIES, STAMEN IN COLD CHILLING NIGHTS, AND EARLY MORNING HOURS, UNDER C. I. S. ORDER THE CUSPIDS CLOSE AT NIGHTS, THE EARLY STAGE PETALS, STAMEN AND OVARIES GET PROTECTION FROM CHILLING COLDS DAMAGES OF NIGHT AND EARLY MORNING TIMES.

AND SHORTLY THEREAFTER, WHEN AT THE DAY, THE SUN RISE AND WEATHER GET WARMER, THE S.I.S. SENSE BETTER TEMPERATURE AND AMPLE LIGHT PARTICLES AROUND AND THEIR DIRECTIONS, THE C. I. S. GET THE S.I.S. REPORTS. ISSUE THE ORDERS FOR OPENING CUSPIDS AND BUDS AGAIN, AS THE SUN SHINE WITH AMPLE DIFFERENT LIGHT PARTICLE FROM PLANET SUN FILLS ALL EARTHS AIR, THERE AFTER C. I. S. ORDERS FLOWERS STAY WIDELY OPENS DURING ALL DAY, EVEN THROUGH THOSE C.I.S. ORDERS, STEMS, FLOWERS, LEAVES, TILTS AND ROTATIONS, CHASE THE DIRECTION OF SUNS TRAVEL PATH, EVERY DAY FROM EAST TO WEST, UNDER WELL FUNCTIONING PLANTS CENTRAL INTELLIGENCE SYSTEMS CENTERS ORDER. AND ACCUMULATING THE NEEDED LIGHT PARTICLES FOR CONSTRUCTIONS OF PLANTS ELECTRONS NUCLEONS PARTICLE COMPOUND CONSTRUCTIONS, THERE AFTER LATE AFTERNOON WHEN SUN GOES DOWN TOWARD THE NIGHT AND TEMPERATURE CHANGING TO COLD ONLY IN SOME SPECIES THE FLOWERS AGAIN UNDER CENTRAL PLANT INTELLIGENCE SYSTEM CENTERS START CLOSING CUSPIDS AGAIN, THESE BIOLOGICAL PLANT MOVEMENT CYCLES CONTINUE FOR LIFE WITH NO INTERRUPTION

SAME DAY AND NIGHT, THIS IS PHENOMENON OF LIGHT FUNDAMENTAL PARTICLE COLLECTING – STORING CYCLES CONTINUE ONE DAYS AFTER THE OTHER, THE COLLECTED Y. F.P. DIRECTLY INTER INTO SUBSYSTEM-UNITS OF DIFFERENT PLANTS ELECTRONS – NUCLEONS AND PRODUCE ELECTRON NEO-GENESIS, NUCLEON NEO- GENESIS, ATOM GENESIS OF PLANT CELLS, THESE PHENOMENONS ALREADY EXPLAINED OTHER CHAPTERS AND VOLUMES AS WELL.

THE FLOWERS WHOSE ATOMS CONSTRUCTED WITH HIGH ELECTRIC ENERGY PARTICLES, PROVIDE BETTER IMMUNITY AND DEFENSE SYSTEM TO PLANTS

EFFECT OF PARTICLE CONSTRUCTIONS ON PLANT IMMUNITY AND DEFENSE SYSTEMS

THE PLANTS CONSTRUCTIONS, THAT HAS BEEN CONSTRUCTED THROUGH USE OF HIGH ENERGY LIGHT PARTICLES COMPOUND CONSTRUCTIONS, AND THOSE WHO HAVE HIGH ELECTRIC ENERGY CONTENT PARTICLE COMPOUNDS, PROVIDE BETTER IMMUNITY AND SURVIVAL TO FLOWERS AND PLANTS.

THE PARTICLES COMPOUND CONSTRUCTED PLANTS, WITH LESS ELECTRIC ENERGIES PARTICLES, PROVIDE LESS IMMUNITY AND DEFENSE CAPABILITIES TO FLOWERS AND PLANTS, THESE PLANT SPECIES CAN NOT DEFEND SELF AGAINST ENVIRONMENTAL DISASTERS, AS THOSE PLANTS, WHO HAVE

BEEN CONSTRUCTED WITH HIGH ENERGY PARTICLE COMPOUND CONSTRUCTED PLANTS.

THE IMMUNITY SYSTEMS OF VIOLET COLOR PLANTS ARE BETTER THAN IMMUNITIES SYSTEMS OF WHITE COLOR FLOWERS. THE VIOLET PLANTS POSSESS HIGH ELECTRIC ENERGY VIOLET FUNDAMENTAL PARTICLES.

BETWEEN SENSIBLE LIGHT PARTICLES IN EARTH, THE VIOLET COLOR LIGHT PARTICLES POSSESS HIGH ELECTRIC ENERGY CONTENT THAN, THE OTHER COLOR SENSIBLE PARTICLES. THE FLOWERS WHOSE ELECTRONS AND NUCLEONS CONSTRUCTED WITH VIOLET COLOR LIGHT PARTICLE COMPOUND CONSTRUCTIONS, THOSE FLOWERS POSSESS BETTER AND HIGHER IMMUNITY AND DEFENSIVE SYSTEMS THAN, THE WHITE COLOR FLOWERS, WHOSE INTERNAL ELECTRON NUCLEON PARTICLE COMPOUND CONSTRUCTIONS ARE DONE BY WHITE COLOR PARTICLE COMPOUND CONSTRUCTIONS.

THE FLOWERS THAT HAS BEEN CONSTRUCTED FROM WHITE COLOR LIGHT PARTICLE –COMPOUNDS OR HAVE BEEN CONSTRUCTED FROM YELLOW COLOR LIGHT PARTICLE COMPOUNDS POSSESS LESS ELECTRIC ENERGY CONTENTS, AND THEIR DEFENSIVE SYSTEMS AND IMMUNITY ARE LESS THAN THE FLOWERS WHO HAVE BEEN CONSTRUCTED FROM VIOLET COLOR LIGHT PARTICLE COMPOUND CONSTRUCTIONS, OR VIOLET COLOR FLOWERS.

VIOLET COLOR FLOWER SPECIES IN HARSH DANGEROUS

ENVIRONMENTAL CONDITIONS SUCH AS EXTREME COLD OR EXTREME HOT WEATHERS HAVE MUCH MORE CHNACE OF SURVIVAL THAN THE WHITE OR YELLOW COLOR FLOWERS, HAVING MORE ELECTRIC ENERGY CONTENT VIOLET COLOR PARTICLE COMPOUNDS PROVIDE BETTER RESISTENCES AND IMMUNITY AND DEFENSE SYSTEMS AGAINST HOT WEATHER AS WELL AS AGAINST COLD WEATHER HAZARDOUS CONDITIONS, WHICH UNDER THOSE LEVEL HEAT OR FREEZING CONDITIONS MOST WHITE OR YELLOW FLOWERS DIE, BUT THE VIOLET FLOWERS AND PLANTS MAY SURVIVE.

PLACING VIOLET COLOR AND WHITE COLOR PLANTS UNDER COLD ENVIRONMENTS, THE VIOLET COLOR FLOWERS SURVIVE BETTER THAN WHITE COLOR FLOWERS, IT IS THE SAME, PLACING VIOLET AND WHITE COLOR FLOWERS UNDER EXTREME HARSH HOT SUMMER TEMPERATURES WITH DEHYDRATIONS, THE VIOLET COLOR PLANTS POSSESS BETER CHANCES OF SURVIVAL THAN THE WHITE COLOR FLOWERS.

HIERARCHY ORDER SYSTEMS BETWEEN S. I. S. AND C. I. S. AND M. I. S. IN PLANTS

THE PLANTS SENSARY INTELLIGENCE SYSTEMS CENTERS (S. I. S.) FROM FLOWERS LEAVES STEMS ROOTS SENSE AND REPORT ENVIRONMENTAL CONDITIONS SUCH AS EXISTING LIGHT PARTICLES AT AIR AND ITS DIRECTIONS AS WELL AS THE OUTSIDE TEMPERATURE CONDITIONS EITHER COLD OR WARM TEMPERATURE, EXISTING WATER OR MINERAL CONDITIONS, ETC., AND REPORT ALL OUTSIDE INFORMATIONS THROUGH Y. - F.P. – I.I. – P.cl. TO PLANTS CENTRAL

INTELLIGENCE SYSTEM CENTERS THROUGH PARTICLE CIRCULATION SYSTEMS, UNDER HIERARCHY ORDER SYSTEMS OF REPORTING INFORMATIONS FROM LOWER INTELLIGENCE SYSTEM CENTERS, TO HIGHER INTELLIGENCE SYSTEM CENTERS.

IN CENTRAL INTELLIGENCE SYSTEM CENTERS (C. I. S.) THE INCOMING F.P.- I.I.- P. cl. FROM SENSARY SYSTEMS, INTER INTO PLANTS INTER ELECTRON-NUCLEON SUBSYSTEM – UNITS CHEMICAL LAB.S, COMBINE WITH INTER ELECTRON- NUCLEON PARTICLE COMPOUNDS, AND STORE INFORMATIONS IMAGE AS RECORDS KEEPING SYSTEMS, PRODUCE DIFFERENT PARTICLE –COMPOUNDS CONSTRUCTING OF THE INTERNAL ELECTRON- NUCLEON OF PLANTS C. I. S., THIS IS PHENOMENON OF NEW-PARTICLE-CLOUD COMPOUND CONSTRUCTION PHENOMENONS, AND THAT IS KEEPING AND STORING OUTSIDE INFOMATIONS AND IMAGES RECORDS, AT INSIDE THE PLANTS INTERNAL ELECTRONS AND NUCLEONS IN PARTICLE COMPOUND FORMS.

THE C.I.S. THROUGH A. S. I. F.P. Mol. C.I.C. COMPARE AND ANALYZE INCOMING PARTICLE CLOUDS INFORMATIONS AND IMAGES WITH PRE-EXISTING INTERNAL ELECTRON-NUCLEON PARTICLE COMPOUND INFORMATIONS AND IMAGES, AND THE C.I.S. THEREAFTER MAKE PROPER DECISIONS OVER THESE INTERNAL ELECTRON NUCLEONS INFORMATIONS AND THEIR COMPARISONS WITH EACH OTHER, THROUGH EXPONENTIAL INTERACTION SPEEDS. THEREAFTER THE C.I.S. ISSUE ORDERS TO BE TRANSMITTED TO LOWER HIERARCHY SYSTEMS, BEFORE THE MOTOR

INTELLECTUAL SYSTEMS CENTERS, THE ORDERS TO BE EXECUTED ACCORDINGLY AS ISSUED BY C.I.S. EXACTLY. UNDER HIERARCHY SYSTEMS FROM HIGHER CENTERS TO LOWER INTELLIGENCE SYSTEM CENTERS, THE M. I. S. EXECUTE ORDERS AS RECEIVED FROM C. I. S. EXACTLY ACCORDING ORDERS INSTRUCTIONS.

FUNDAMENTAL PARTICLE WEAPONS OF MASS DESTRUCTION, AND CRIMES AGAINST HUMANITY SUPER-SONIC PARTICLE WEAPONS WOUNDS AND INJURIES MICROWAVE WEAPONS OF MASS DESTRUCTION AND MICROWAVE WEAPON WOUNDS, THE CRIMES AGAINST HUMANITY

SHOOTING MICROWAVE SUPER-SONIC LIGHT CODED LASER -PARTICLES WEAPONS OF MASS DESTRUCTION, CAN BE FIRED AIRBORN FROM ANY DISTANCE, FROM ROUGES ARMY BASE, OR FROM POWERFUL TELECOMMUNICATION TOWERS, THROUGH COMPUTERIZED WELL CALIBRATED AND DOSE ADJUSTED VARIABLE LETHAL AND NON-LETHAL MICROWAVE DOSES DELIVERED COMPUTERIZED INTO ANY SELECTED HOUSES IN ANY CITY. THE ROUGES AGENTS CAN SHOOT OR TORTURE BY MICROWAVE ANYONE INSIDE THEIR OWN HOMES WITHOUT LEAVING ANY TRACE, THE POLICE IS PART OF THE GANG. USE OF THESE DEVICES DO NOT LEAVE TRACE FROM KILLING IN BEHIND.

MICROWAVE SUPERSONIC WEAPONS INJURIES ARE INTERNAL ELECTRON-NUCLEON PARTICLE COMPOUND CONSTRUCTIONS, THE PARA- CLINICAL STUDIES SUCH AS CHEMICAL STUDIES, MRI, CAT SCANS, PET SCAN OR SIMILAR

STUDIES MAY NOT HELP? WE MAY HAVE SOME IN THE FUTURE, THESE ABUSES ARE COMMON IN LESS ADVANCED REGIONS AT MOST.

THE INTERESTING FACT IS, THAT INTERNAL ATOM INJURIES BY MICROWAVE PRODUCE PATIENT'S COMPLAINTS, SIMILAR TO COMPLAINTS OF INJURIES CAUSED BY OTHER REASONS AND DISEASES, AND THE HISTORY TRACE IS SOME THING SIMILAR AS ABOVE, BUT THERE IS NO VISIBLE PHYSICAL EVIDENCES.

DESTRUCTION OF INTERNAL ELECTRON NUCLEON S. FP - I.I. - P.cl. COMP. CONSTRUCTION BY MICROWAVE

PATHOLOGY OF MICROWAVE –LASER INJURIES

PATHOLOGY OF PERMANENT NON STOP TINITUS IN PATIENTS WITH MICROWAVE INJURY

POWERFUL DESTRUCTIVE BIOCIDAL SUPERSONIC MICROWAVE FUNDAMENTAL PARTICLES INTER INTO INSIDE CNS AUDITARY CENTERS INTERNAL ELECTRONS- NUCLEONS SUBSYSTEM UNITS CHEMICAL LAB.S AND DISRUPT FINE S. Y. – F.P. – I.I. – P.cl. _ COMPOUNDS CONSTRUCTIONS IN CNS ELECTRONS -NUCLEONS, MOST OF THESE PARTICLE-CLOUD COMPOUND CONSTRUCTIONS ARE IMPORTANT FACTORS AT PRODUCTION OF THOUGHT CURRENTS WHICH ARE S. Y. – F.P. I.I. – P.cl. CURRENTS OF THE DIFFERENT KINDS CONSTRUCTED THROUGH DIFFERENT NORMAL Y. S.- F.P.- I.I.- P.cl. –COMPOUND COSTRUCTIONS HISTORY OF MAN KIND

DURING PAST HUMAN LIFE HISTORY,

INTERANCE OF DESTRUCTIVE BIOHOSTILE POWERFUL MICROWAVE FUNDAMENTAL PARTICLE INTO CNS- ELECTRONS AND NUCLEONS UPON INTERANCE TEAR APART AND BREAK DOWN MOST OF FINELY PRE-CONSTRUCTED BIOFRIENDLY PRE-EXISTING PREVIOUSLY CONSTRUCTED Y. S. –F.P. –I.I. – P.cl. _ COMPOUND CONSTRUCTIONS OF CNS ELECTRONS - NUCLEONS,

IN CONTINUATION OF CNS ELECTRON-NUCLEON PARTICLE COMPOUND DESTRUCTIONS, THE HARSH BIO-CIDAL MICROWAVE PARTICLES COMBINE WITH REMNANTS OF INTER ELECTRON- NUCLEON PARTICLE COMPOUNDS UNDER REGENERATIVE OR DEGENERATIVE A. S. I. F.P. Mol. C.I.C. AND PRODUCE NEWLY CONSTRUCTED NON-LIVE MICROWAVE PARTICLE-COMPOUNDS CONSTRUCTIONS INSIDE CNS AUDITARY CENTERS ELECTRON-NUCLEONS,

THE PRODUCED ROARING MICROWAVE PARTICLE COMPOUND CONSTRUCTIONS INSIDE CNS ELECTRONS –NUCLEONS CAUSE CREATIONS OF PERSISTENT HARSH ROARING TINITUS IN PATIENTS HEAD WHICH THESE TINITUS ARE NON INTERRUPTIVE AND PERMANET TINITUS WITH NO RELIEF, AND RELATE TO EXISTENCES OF MICROWAVE PARTICLES COMPOUNDS IN CNS ELECTRONS-NUCLEONS PARTICLE COMPOUND CONSTRUCTIONS,

THE PHENOMENON OF NON-LIVE BIOCIDAL MICROWAVE PARTICLE-COMPOUND NEO-GENESIS AND CONSTRUCTIONS

OF CNS- ELECTRONS AND NUCLEONS PARTICLE COMPOUND WITH MICROWAVE PARTICLE CONSTRUCTIONS ARE MAIN CAUSE OF PRODUCTION OF PERSISTING HARSH NON STOP TINITUS SECONDARY TO EXISTING MICROWAVE PARTICLE COMPOUNDS INSIDE BRAIN ATOMS WHICH PRODUCES NON-STOP TINITUS INSIDE HEAD WITH NO RELIEF,

PHYSIOLOGICAL FUNCTIONS OF CERUMEN

WHEN PATIENT'S HEAD SHOT BY WEAPON GRADE MICROWAVE FROM ANY DISTANCES, THE MICROWAVE -LASER PARTICLES INTER INTO HEADTHROUGH EXTERNAL EAR CANALS, AND THROUGH EXTERNAL EAR CANAL, MIDDLE EAR, AND INNER EAR TRANSIT DIRECT INTO CNS AUDITARY CENTER INTERNAL ELECTRON-NUCLEON STRUCTURES, POWERFUL MICROWAVE DESTRUCT EVERY STRUCTURE IN THE PATH, DEPENDING WHAT KIND AND WHAT DOSE, MICROWAVE SHOT INTO HEAD, AND HOW MUCH AND HOW LONG CONTINUOUSLY RUNNING INTO BRAIN.

AT EXTERNAL EAR CANAL THE COPIOUS EXCRETION OF THE CERUMEN IS THE MOST IMPORTANT DEFENSE SYSTEMS TO DIMINISH THE BIOHOSTILE SONIC PARTICLES DESTRUCTIONS, CERUMEN'S MAIN JOB IS NOT THE INSECT REPELLENT. IN ORDER TO DIMINSH HARSH POWERFUL DESTRUCTIVE VIBRATIONS AND ROARING POWERFUL CURRENTS OF MICROWAVE PARTICLE LASER CURRENTS, AND PROTECT MIDDLE EARS FINE BONY STRUCTURES FROM DISRUPTIONS ,AND TYMPANIC MEMBRANE RUPTURE, INJURIES OF CNS AND INNER EAR NEURAL CONSTRUCTIONS IS

COPIOUS CERUMEN EXCRETION BY EXTERNAL EAR, WHICH COATS ALL AROUND THE EXTERNAL EAR CANAL WITH INSTANT EXCRETIONS AND COVER T.M. AS WELL, AND HELP TO REDUCTIONS OF DESTRUCTIVE POWERS OF THE MICROWAVE.THIS IS THE MAIN FUNCTION AND ESSENTIAL PHYSIOLOGICAL FUNCTIONS OF EAR SERUMEN EXCRETIONS.

THE EXTERNAL EAR GLANDS EXCRETE COPIOUS QUANTITIES OF CERUMEN AND COVER THE T. M. AND EXTERNAL EAR CANANL LUMEN ALL AROUND , AND DIMINISH ROARING HARSH DESTRUCTIVE SHEARING FREQUENCIES, AND COMBAT BIOHOSTILE MICROWAVE VICIOUS SONIC − FUNDAMENTAL PARTICLES WHICH TEAR APART ANY FINE BIOLOGICAL TISSUES IN ITS PATH, THIS PHENOMENON OF CERUMEN EXCRETIONS AND EXISTENCE IN EAR CANAL IS CREATED TO COMBAT AND CONTROL BIOHOSTILE HARSH SONIC PARTICLE INJURIES AND PROTECT SENSITIVE VITAL STRUCTUTRES OF CNS AUDITARY SYSTEMS,

THE EXCRETED CERUMEN COVER LUMENS OF EXTERNAL EAR CANAL AND TYMPANIC MEMBRANE, TO DIMINISH DESTRUCTIVE FORCES OF BIOHOSTILE INDUSTRIAL MICROWAVE PARTICLE CURRENTS, TO PREVENT SHEARING ACTIONS OF HARSH VIBRATING FREQUENCIES OF MICROWAVE PARTICLES IN ORDER THE T.M., FINE EARS BONY STRUCTURES AND FINE NEURAL TISSUES OF INNER EAR AND BRAIN NOT TORN APART INTO PIECES BY INDUSTRIAL MICROWAVE FORCES, THE CERUMENS MAIN BIOLOGICAL FUNCTIONS IS COMBATING AGAINST BIOHOSTILE SONIC AND MICROWAVE PARTICLES IN ORDER PREVENT THEIR

DESTRUCTIONS OF AUDITARY CONSTRUCTIONS,

THE WAR SYNDROMES

SYMPTOMS AND SIGNS OF MICROWAVE
WEAPONS BRAIN-INJURIES

SYMPTOMS AND SIGNS OF PATIENTS WHO HAS BEEN SHOTS BY MICROWAVE WEAPONS OF MASS DESTRUCTION ON HEAD, AND BRAIN INJURIES, CHRONIC DESTRUCTION OF CNS ELECTRONS NUCLEONS:

WHEN DURATION OF BIOHOSTILE POWERFUL MICROWAVE SHOTS ON HEAD CONTINUES FOR HOURS OR MORE, PATIENT BECOME RESTLESS, THE POTENT HIGH POWER INDUSTRIAL BIOCIDAL LASER- MICROWAVE SHOTS ON HEAD OF INDIVIDUALS WITH BORDER LINE HEALTH CONDITIONS PRODUCE CONTINUOUS STIFF NECK, NAUCEA, VOMITING HEADACHES SIMILAR TO MENINGEAL IRRITATION SIGNS AND SYMPTOMS, BUT IN PARA-CLINICAL TESTS THE FINDINGS ARE NON- CONCLUSIVE AND NORMAL LOOKING, MICROWAVE DAMAGES ARE INTERNAL ELECTRON-NUCLEON INJURIES BY OBNOXIOUS BIOHOSTILE PARTICLES BUT MENINGITIS PRODUCED BY MICRO-ORGANISM OR OTHER SIMILAR CAUSES WITH CELLS AND TISSUE DAMAGE INVOLVEMENTS.

THE INDIVIDUALS WHO ARE YOUNG, STRONG, HEALTHY WITH GOOD BODY IMMUNITY, RESIST BEING KILLED BY

MICROWAVE INSULTS.

BUT MOST WITHOUT EXCEPTIONS SUFFER FROM VARIABLE DIFFERENT MICROWAVE CHRONIC COMPLICATIONS, AND PERMANENT DISABILITIES SUCH AS NERVE DEAFNESS, PERMANENT NON - STOP TINITUS, AND CHRONIC DIFFERENT DEGREES BRAIN DAMAGES.

THE SYMPTOMS AND COMPLAINTS THAT CAUSED BY MICROWAVE INJURIES, ARE SIMILAR TO THE SYMPTOMS AND COMPLAINTS WHICH PRODUCED THROUGH THE OTHER DISEASES OF BRAIN DAMGE, LIKE: MENINGEAL INFECTIONS TYPE COMPLAINTS, GUN SHOT PRODUCED SYMPTOMS, BLUNT INJURIES TYPE COMPLAINT, ETC.).

GENERALLY, MICROWAVE DOE'S NOT LEAVE DEMONSTRABLE SIGNS TO PROVE THE EVIDENCE OF MICROWAVE INJURIES IN BEHIND, NO CUTS SIGNS, NO MASS OR REDNESS SIGNS, ETC. THE INJURED BRAIN BY MICROWAVE NEED PROLOGED PROPHYLAXIS ANTIBIOTIC TRAETMENTS, BECAUSE INJURED TISSUE BY MICROVAWE IS SIMILAR TO RADIATION WOUND, DO NOT HEAL FAST, MICROWAVE DAMAGES MOSTLY ARE PRONE TO SECONDARY INFECTIONS ETC. ALSO MICROWAVE INJURIES HEALS VERY SLOWLY OVER YEARS TO RECONSTRUCT ELECTRONS NUCLEONS CONSTRUCTIONS, ALSO THE HEALING PROCESS MAY NEVER OCCUR AS WELL AFTER LONG TIMES IN SOME CASES.

WHEN PATIENTS SHOT BY MICROWAVE, ALWAYS PATIENTS FEEL TRANSITION PATHS OF HARSH POWERFUL ROAMING

MICROWAVE CURRENTS, PASSING THROUGH EXTERNAL EAR CANAL TOWARD THE AUDITARY CNS CENTERS, AND THE PATIENTS CAN SHOW WITH FINGER POINT THE PLACE OF PAIN OVER SKULL, WHICH IT CORRELATES TO CNS HEARING CENTER LOCATION AT BRAIN.

THE MICROWAVE DEAFNESS AND HARSH CONTINUOUS TINITUS ROARS WITH NO STOP ALWAYS IS THE FIRST COMPLAINT THE PATIENT STATES WHEN SHOT BY MICROWAVE ON HEAD. PATIENTS MOSTLY IN SERIOUS INJURIES MANIFEST BRAIN INJURY SIGNS AND SYMPTOMS.

THE SYMPTONS IN SERIOUS CASES ARE HEADACHE, STIFF NECK, NAUSIA, VOMITING, CONFUSION, FEVER, REMARKABLE NECK RIGIDITY, GRADUAL HEARING LOSS, CONTINUOUS HARSH ROARING WITH NO STOP TWENTY - FOUR HOURS, ETC. BUT CHANGES IN XRAYS, LAB. TESTS NON SPECIFIC, MAY NOT HELP. THE MRI MAY NOT HELP TO DIAGNOSIS.

FACT ABOUT INTERNATIONAL
MICROWAVE PARTICLE WARS

IN MODERN INTERNATIONAL WARS, SHOOTING WITH MICROWAVE FUNDAMENTAL PARTICLE'S CODED WITH LIGHT – LASER WEAPONS (BULLETS). OR UNSING ANY OTHER BIOHOSTILE PARTICLES, THESE WEAPONS CONTAMINATE AIR AND SPACE WITH OBNOXIOUS PARTICLE CLOUDS. AND THESE ARE WEAPONS OF MASS DESTRUCTIONS AND CRIMES AGAINST HUMANITY THAT DAMAGES ORDINARY

PEOPLES, CHILDREN, WOMEN, MEN ALIKE AS THE SOLDIERS ARE NOT AWARE ABOUT THE PARTICLES IN AIR, THAT THEY ARE LIVING AND BREATHING.

WHEN AIRBORN MICROWAVE- LASER PARTICLES INTER INTO AUDITARY CNS CENTERS INTERNAL ELECTRON-NUCLEON, DESTRUCT AND TEAR DOWN NORMAL S.Y. – F.P. – I.I. - PARTICLE CLOUD –COMPOUND CONSTRUCTIONS INSIDE BRAINS ELECTRONS –NUCLEONS. THE CNS INTERNAL ELECTRON-NUCLEON S. Y. – F.P. – I.I.- P. cl. _ COMPOUND DESTRUCTIONS.THE SYMPTOMS AND SIGNS ARE SIMILAR TO THE THOSE SYMPTOMS AND SIGNS OF NEUROLOGICAL BRAIN INJURY DISEASES.

THE SOLDIERS AND NATIVE CITIZENS MAY SUFFER FROM ACUTE PARTICLE INJURIES SYNDROMES WHEN THEY ARE IN NEIBORHOOD OR IN FIELD, OR THEY MAY DEVELOPE CHRONIC COMPLICATIONS IN LATER TIMES.

PARTICLE SHOTS INTO CHEST CAVITY AND INTER- ABDOMINAL ORGANS

PATTERNS OF MICROWAVE LASER FUNDAMENTAL PARTICLES TRANSIT INSIDE BODY TISSUES

INSIDE TISSUES FUNDAMENTAL PARTICLES CURRENTS FLOW, IN STAIGHT LINES, AS SEPARATE PARALLEL PARTICLE CURRENTS, PARALLEL PARTICLE CURRENTS FLOW SEPARATE

AND APART FROM EACH OTHER.

INSIDE CHEST CAVITY BIOHOSTILE POWERFUL MICROWAVE PARTICLES DO NOT FLOW AS ONE SINGLE PARTICLE CURRENT FROM ONE SIDE TO OPPOSITE POINT, THE POWERFUL OBNOXIOUS MICROWAVE PARTICLES FLOW INSIDE THE CHEST AND ABDOMINAL CAVITY AS: "S. L. – S. P. – P. C." SEPARATE PARTICLE CURRENTS FROM EACH OTHER, IN MULTI-STRAIGHT LINE SEPARATE INDEPENDENT FLOWS.

CONTRARY TO BIO-NEUTRAL MICROWAVE FUNDAMENTAL PARTICLES WHICH, ALL ARE NON SENSIBLE. IN CONTRAST THE BIOHOSTILE MICROWAVE INTERNAL BODY PARTICLE CURRENTS OR FLOWS, ALL ARE SENSIBLE WHEN CROSSING INSIDE THE HEART, LUNG, LIVER, AND OTHERS.

THIS SENSIBILITIES OF INDUSTRIAL POWER BIOHOSTILE MICROWAVE PARTICLES, THROUGH SENSING THE PARTICLES REVEALS, THAT THE BIOHOSTILE MICROWAVE PARTICLES CURRENTS ARE NOT SINGLE ONE STREAM PARTICLE FLOWS INSIDE BODY TISSUES, BUT BIOHOSTILE MICROWAVE PARTICLES CURRENTS FLOW THROUGH: MULTIPLE DIFFERENT PARALLEL CURRENTS ONE NEXT TO OTHER IN STRAIGHT LINES BETWEEN INTERANCE POINT TO CHEST, BELLY AND BODY, AND STAY SENSIBLE THROUGH OUT ALL INTERRNAL BODY TRANSIT TIME, ALL INTERNAL BODY TRAVEL PATH, UNTIL IT EXIT FROM OTHERSIDE OF BODY THROUGH EXIT POINT, ALL ARE SENSIBLE.

THE HUMAN SPECIES WILL SENSE PAINLESSLY THE

BIOHOSTILE PARTICLES, WHEN THESE MICROWAVE PARTICLE CURRENTS SHOTS INTO HUMAN'S CHEST AND ABDOMEN AND PARTICLE CURRENTS TRAVELING INSIDE DIFFERENT THORACO-ABDOMINAL TISSUES ALL ARE CLEARLY SENSIBLE.

WHEN THE BIOHOSTILE MICROWAVE PARTICLE CURRENTS INTER INTO THE CHEST CAVITY, AND FLOWING AS SEPARATE MULTIPLE PARALLEL PARTICLE CURRENT FLOWS, IN STRAIGHT LINES FROM POINT OF INTERANCE TO THE CHEST WALL DIRECT TO THE EXIT POINT OF BIOHOSTILE PARTICLES EXIT POINT FROM OTHERSIDE TO OUTSIDE BODY.

THE HUMAN BODY WILL SENSE PAINLESSLY THE FLOW OF BIOHOSTILE PARTICLE CURRRENTS INSIDE DIFFERENT INTERNAL BODY ORGANS SUCH AS: HEART, LUNGS, LIVER, AND INSIDE ABDOMINAL ORGANS WHEN THE MICROWAVE CURRENTS FLOWING AND CROSSING DIFFERENT BODY TISSUES. FINALLY EXITING FROM THE OPPOSITE SIDES OF ENTERANCES TO OUTSIDE.

THORACO- ABDOMINAL INJURIES BY
SUPERSONIC MICROWAVE PARTICLES

BIOHOSTILE MICROWAVE PARTICLES CURRENTS
INTERACTIONS WITH BODY ORGANS – PCS:

THE POWERFUL PARALLEL, SEPARATE PARTICLE CURRENTS

OF MICROWAVE SUPERSONIC FUNDAMENTAL PARTICLES INSIDE THE CHEST CAVITY ALL FLOW IN STRAIGHT LINES, AS SEPARATE CURRENTS PARALLEL TO EACH OTHER, BETWEEN THE ENTERANCE POINT OF THORACIC WALL IN ONE SIDE OF THORAX, DIRECT LINE FLOW TO THE EXIST POINT IN THE OPPOSITE SIDE OF THE THORACIC CAGE.

THE MICROWAVE PARTICLE FLOW INSIDE THE THORACIC CAVITY TISSUES ARE MULTIPLE PARALLEL, SEPARATE FLOWS, ALL DIFFERENT PARALLEL CURRENTS CROSS AND INTER INTO DIFFERENT INTER –THORACIC OR INTER-ABDOMINAL TISSUE IN STRAIGHT LINES FROM ONE POINT INTERANCE TO OPPOSITE SIDE OF EXIT, THESE DIFFERENT PARALLEL STRAIGHT LINE PARTICLE FLOWS DO NOT JOIN TOGETHER, AND DO NOT FLOW AS ONE LARGE SINGLE STREAM FLOW, OR SINGLE ONE LARGE CURRENT IN THORACIC CAVITY TISSUES. THIS PATTERN OF PARTICLE CURRENTS ALL SENSED CLEARLY BY VICTIMS. THESE ARE NUMEROUSLY SENSED FACTS AT OVER HUNDREDS EPISODES.

AS BEFORE EXPLAINED, THE ALL ANIMALS, LIVING THINGS ORGAN'S FUNCTION OPERATES UNDER THEIR ORGAN-SPECIFIC PARTICLE INTELLIGENCE SYSTEM CENTERS, AND ALL ELECTRONS NUCLEONS POPULATIONS OF THE ORGAN IS CONNECTED TO EACH OTHER, THROUGH ORGAN SPECIFIC PARTICLE CIRCULATION SYSTEMS.

WHEN THE POWERFUL LETHAL DOSES BIOHOSTILE LIGHT LASER CODED MICROWAVE SHOTS INTERING INTO DIFFERENT THORACO-ABDOMINAL ORGANS AND SYSTEMS,

DEFFINITELY THESE BIOHOSTILE PARTICLE CURRENTS WILL INTERFERE WITH THESE VITAL ORGANS PARTICLE CIRCULATION SYSTEMS (PCS), SPECIFICLY IN ELDERLY, AND BORDERLINE HEALTH CONDITION INDIVIDUALS.

IN ELDERLY AND BORDER LINE HEALTH CONDITIONS, THE BIOHOSTILE MICROWAVE PARTICLES PRODUCE SERIOUS IRREVERSIBLE MEDICAL COMPLICATIONS SUCH AS DYSRHYTHMIA, WHEN BIOHOSTILE LASER MICROWAVE PARTICLES INTER INTO HEART TISSUES AND CROSS CARDIAC ORGAN, CAN DISRRUPT CARDIAC ELECTRICAL PARTICLE CIRCULATION SYSTEMS OF THE HESS- INTELLIGENCE SYSTEM CENTER, THE HESS INTELLIGENCE SYSTEM CENTER THROUGH CARDIAC PARTICLE CIRCULATION SYSTEMS (P.C.S.) CONTROLS ALL INTER -CARDIAC ELECTRONS NUCLEONS TOTAL POPULATIONS NANO – FUNCTIONS. THE POWERFUL STRONG BIOHOSTILE MICROWAVE LASER PARTICLES DISRUPT CARDIAC PARTICLES CIRCULATIONS SYSTEMS FROM HESS CENTRAL INTELLIGENCE SYSTEM CENTERS, OR FROM ANY OTHER MAJOR BRANCHES TO ELECTRICAL FUNDAMENTAL PARTICLE CIRCULATION SYSTEMS OF THE HEART, ANY DISRUPTION OF HESS – P. C. S. CAN CAUSE CARDIAC FUNCTION DISORDERS AND COMPLICATIONS SUCH AS: DYSRHYTHMIA, C. H. F., M. I., CARDIAC ARREST, AND DEATHS.

THE PHENOMENONS SIMILAR TO HEART MAY OCCUR, IN OTHER INTRA-THORACIC ORGANS, IN LUNGS, AND OTHERS, UNDER LETHAL DOSES OF BIOHOSTILE MICROWAVE PARTICLES SHOTS. MICROWAVE EASILY CAN TEAR DOWN WHOLE OPERATTIONS OF DIFFERENT BODY ORGANS AND SYSTEMS

PARTICLE CIRCULATIONS SYSTEMS TOTAL OPERATIONS IN SECONDS, AND CAUSE ALL ORAGANS AND SYSTEMS TOTAL ELECTRONS NUCLEONS POPULATIONS NANO-FUNCTIONS DISAPPEAR, THAT MEANS NO BIOLOGICAL EXISTANCE AND FUNCTIONS AND DEATH.

THE MICROWAVE INJURIES BY
MICROWAVE WEAPONS IN WARS

MICROWAVE USE IN SOME NATIONS AGAINST CITIZENS

IN THE CASES OF BIOHOSTILE STRONG LASER - MICROWAVE SHOT - INJURIES ON HEAD WHEN SHOOTING DIRECT ON THE HEAD CAUSE IRREPAIRABLE DAMAGES FREQUENTLY PRODUCE HEADACHES, STIFF NECK, NAUSIA LOCAL TENDERNESS EXACTLY ON SKULL SIDE OVER THE INJURED CNS- ELECTRONS- NUCLEONS INJURED TISSUES AND LOW LEVEL FEVER ETC. ALL ARE DETECTABLE BY EXAMINERS,

IN CONTRAST THE CHEST INJURIES DOES NOT PRODUCE TOO MUCH SIGNS OR SYMPTOMS IN NORMAL STRONG INDIVIDUALS, UNLESS THE OCCURANCES OF COMPLICATIONS SUCH AS SUDDEN MI, CHF, CARDIO-PULMONARY ARREST OR SIMILAR SITUATIONS AND DEATH WHEN OCCUR SPECIFICLY IN BORDERLINE HEALTH INDIVIDUALS WHEN HIT BY POWERFUL INDUSTRIAL SUPRASONIC PARTICLE CURRENTS ON HEART AND VITAL ORGANS THROUGH PRODUCTION OF DYSRHYTHMIAS, BLOCKING CARDIAC RHYTHM

CONDUCTIONS AND PRODUCING HEART –ARREST SECONDARILY CAUSING M.I., ETC., MOSTLY OCCUR IN BORDERLINE HEALTH INDIVIDUALS, MOSTLY PRONOUNCED DEAD BY NATURAL CAUSES AND CERTIFICATES ISSUED ROUTINELY. NOBODY CAN TALK AGAINST.

MICROWAVE SHOT OF BORDERLINE INDIVIDUALS EASILY CAN PRODUCE AND PROVOKE DYSRHYTHMIC CARDIOPULMONARY MALFUNCTIONS AND DEATH, CARDIAC ARREST, RHYTHM FAILURE, AND SECONDARILY ALL OF ABOVE PRODUCE DETECTABLE DISEASES SUCH AS MI, OTHER CARDIOPULMONARY INCIDENTS, WHICH THERE IS NO DIFFERENCES BETWEEN THESE LASER MICROWAVE DEATHS AND NATURALLY OCCURING DEATHS. ALL LOOK SIMILAR, THIS PHENOMENON ARE ABUSED EXTENSIVELY BY SOME AUTHORITIES AGAINST PEOPLE. IN SOCIALIST BACKED NATIONS.

ROUGES OF SOCIALIST STYLE OPERATED NATIONS, ABUSE WEAPONIZED MICROWAVE STOCK PILES OF COUNTRY'S WEAPONS OF MASS DESTRUCTION AGAINST HUMANITY, IN THE HANDS OF WELL PAID AND TRAINED AGENTS IN OPERATION OF THESE MASS MURDERS, IN BEHIND THE MASSIVE FALSE RADIO- TV PROPAGANDA SYSTEMS, UNDER THESE OPERATIONS, MASS TORTURES AND MASS MURDERS INSIDE THE VICTIMS HOMES THROUGH WELL TRAINED AND PAID AGENTS TO COMMIT MURDERS AND TAKE ONE HALF OF PROPERTIES AS ADDITIONAL REWARDS, IN HUNGRY INFESTED NATIONS, THIS IS THE BEST JOB AND EVERYONE APPLY AND COMPETES FOR JOB. THESE ARE DAILY ORDINARY

AFFAIR BUT YOU MUST NOT EXPOSE. AND THOSE SURVIVE WHO DENIY THESE OUTLAW SYSTEM'S EXISTANCE. THESE ARE THE RULES ORDERED AND YOU MUST FOLLOW THE ORDERS, IF YOU NOT, OKEY, IT IS YOUR TURN IN MICROWAVE'S NEXT LINE, MUST BE ANNIHILATED AS THE NEXT BY AGENTS WEAPONS AND THEY ARE NEXT OWNER OF YOUR HOME. THEN DAILY SHOTS ON YOUR FORHEAD AND THESE ARE ORDERS AND RULES.

YOUNG MAY SURVIVE WITH CHRONIC COMPLICATIONS, BUT BORDER LINE HEALTH INDIVIDUALS AND ELEDERLY INSIDE OWN RESIDENCE GET **NATURAL LOOKING ANNIHILATION WITHOUT LEAVING ANY TRACE FROM MURDER, AND RULERS AFTER MURDER ARRANGE FUNERAL EVEN PARTICIPATE. AND FORENSIC OFFICIAL ISSUE LEGALLY STAMPPED DEATH CERTIFICATES. TODAY SUPRASONIC WEAPONIZED MICROWAVE SHOT KILLINGS IN HANDS OF ROUGE RULERS ARE A GIFT FOR THEM, TO KILL AND SCAPE FROM MURDER, WITHOUT LEAVING A TRACE.

BIOHOSTILE LASER MICROWAVE WEAPONS, AND CHEMICAL, NUCLEAR, BIOLOGICAL WEAPONS ALL ARE WEAPONS OF MASS DESTRUCTION. THIS BOOK IS AN APPEAL TO ILLEGALIZE ALL MICROWAVE WEAPONS OF MASS DESTRUCTIONS, AND ALL BE OUTLAWED ALL OVER THE WORLD, FOR SAFETY OF WORLD PEOPLE.

THERAPY OF MICROWAVE INJURIES

PATHOLOGY OF CNS ELECTRONS – NUCLEONS DAMAGES BY MICROWAVE

1 - THE MAIN PATHOLOGICAL DAMAGES OF WEAPONIZED MICROWAVE INJURIES INTO AUDITARY CNS ELECTONS-NUCLEONS PARTICLE COMPOUNDS CONSTRUCTIONS, ARE DISRUPTION AND DESTRUCTIONS OF: CNS S. F.P. – I.I. – P. cl. _ COMPOUND MOLECULAR CONSTRUCTIONS IN AUDITARY CNS ELECTRONS AND NUCLEONS. THEREAFTER BIOHOSTILE MICROWAVE LASER FUNDAMENTAL PARTICLES DESTROY AUDITARY CNS ELECTRONS NUCLEONS NORMAL S. F.P. – I.I. – P.cl. COMPOUNDS.

THE POWERFUL BIOHOSTILE MICROWAVE PARTICLES TEAR DOWN NORMAL S. –F.P. – I.I. – P. cl. – COMPOUNDS OF CNS AUDITARY ELECTRONS NUCLEONS COMPOUNDS. AND IN THEIR PLACE, NON-LIVE BIOHOSTILE MICROWAVE PARTICLE COMPOUNDS CONSTRUCT HARMFUL PATOLOGICAL MOLECULAR CONSTRUCTIONS. WHICH HARSH CONTINUOUS ROAR WITH NO INTRUPTIONS IS THEIR RESULTS.

THE NORMAL ELECTRON NUCLEON S. –F.P. – I.I. – P. cl. COMPOUND CONSTRUCTIONS OF AUDITARY CNS CENTER, FORCEFULLY TORE DOWN AND REMOVED, THEREAFTER REPLACED BY NON-LIVE IMPLANTED DEAD NON -FUNCTIONAL ROARING MICROWAVE PARTICLE-COMPOUNDS. WHICH IT'S COMPLICATION IS UNABLE TO HEAR AND UNDERSTAND, AS WELL AS NON- INTERRUPTED CONTINUOUS

TINNITUS WITH NO STOP.

2 - THE THERAPY FOR THIS PROBLEM IS ELIMINATION AND REMOVING ROARING DESTRUCTIVE MICROWAVE PARTICLE-COMPOUNDS OUT OF AUDITARY CNS INTER CNS ELECTRONS - NUCLEONS CONSTRUCTIONS, THROUGH DEGENERATIVE A. S. I. F.P. Mol. C.I.C.

IN THEIR PLACE AGAIN CONSTRUCTING AND RE-IMPLANTING NORMAL S. –F.P. – I.I. – P. cl. _COMPOUNDS THROUGH REGENERATIVE A. S. I. F.P. – Mol. C.I.C. INSIDE THE AUDITARY CNS INTERNAL ELECTRONS- NUCLEON PARTICLE COMPOUND, THIS IS RE- INSTALLATIONS AND CONSTRUCTION CNS ELECTRONS NUCLEONS PARTICLE COMPOUNDS WITH NORMAL S. – F.P. – I.I.- P. cl. COMPOUNDS. WITH ANOTHER SIMILAR NEW S. F.P. – I.I. – P. cl. COMPOUNDS AS BEFORE, IN THE PLACE OF PREVIOUS ROARING MICROWAVE COMPOUND CONSTRUCTIONS.

TRYING TO STORE NORMAL PARTICLE CLOUDS, AND PRODUCE NORMAL PARTICLE CLOUD STORAGE AND RETRIEVALS INSIDE THE CNS ELECTRONS NUCLEONS WITH NORMAL BIOLOGICAL PARTICLE COMPOUNDS.

REVERCE PREVIOUSLY PRODUCED NEURAL DEAFNESS, AS WELL AS WITH REMOVAL OF HARSH MICROWAVE FROM CNS ELECTRONS NUCLEONS COMPOUNDS, THE HARSH CONTINUOUS TINNITUS ALSO GRADUALLY WILL ELEIMINATE AND DISAPPEAR.

3 – ABOVE IS TREATMENT OF CHOICE FOR MICROWAVE INJURIES OF CNS INTERNAL ELECTRON NUCLEON REPAIRES. THIS THERAPY WILL BE EXPLAINED IN SECOND VOLUME OF THIS BOOK.

4- THE PARTICLE CLOUD INFECTED DISEASES (OR PSYCHIATRIC DISEASES): WITH ABNORMAL SONIC, LIGHT, ELECTRIC, THERMAL FUNDAMENTAL PARTICLE INFORMATION IMAGE PARTICLE CLOUD TRANSMITIONS AND INFESTATIONS (S. E. Y. T. - F.P. – I. I. – P. cl.), UNDER REGENERATIVE A. S. I. F. P. Mol. C.I.C. THE ABNORNAL PARTICLE CLOUDS TRANSMIT FROM ABNORMAL PARTICLE INFECTED INDIVIDUALS TO NORMAL INDIVIDUAL TRANSMISSIONS, AND PRODUCE OR INFECT WITH ABNORMAL PARTICLE CLOUD INFECTIONS (PSYCHIATRIC DISEASES) THE NORMAL INDIVIDUALS ELECTRON NUCLEONS INTERNAL PARTICLE COMPOUND COMINATIONS AND INFLICTIONS TO ABNORMAL PARTICLE CLOUD INFECTED DISEASES. THE PATIENT'S CNS INTERNAL ELECTRON-NUCLEONS PARTICLE COMPOUND CONSTRUTIONS HAVE BEEN CONSTRUCTED, AND INFLICTED BY ABNORMAL S. Y. E.– I.I. – F.P. – P. cl., WHICH THIS PHENOMENON IS STRORAGE OF THE CNS ELECTRONS NUCLEONS WITH ABNORMAL PARTICLE CLOUDS. OR GETTING INFECTED WITH ABNORMAL PARTICLE CLOUD INFECTIONS AND STORAGE OF DISEASED PARTICLE CLOUDS INCIDE ELECTRON NUCLEONS.

THE ESSENTIAL AND MAIN CAUSE OF PATIENTS WITH PARTICLE INFECTED DISEASES IS STORAGE OF AND ACCUMULATIONS AND INFECTION OF THE CNS ELECTRONS NUCLEONS WITH ABNORMAL PARTICLE CLOUDS, AND TREATMENT OF

CHOICE FOR THIS ABNORMAL INFECTED PARTICLE CLOUD CONSTRUCTIONS STORAGE AND ACCUMULATIONS FROM INFECTED INTERNAL ELECTRON NUCLEONS PARTICLE COMPOUNDS OF CNS ELECTRONS NUCLEONS IS REMOVAL AND ELEIMINATION OF THE ABNORMAL S. Y. E. T. – F.P. – I.I. –P. cl. TO OUT OF THE INFECTED CNS ELECTRONS AND NUCLEONS UNDER DEGENERATIVE A. S. I. F.P. Mol. C.I.C.

THIS IS PHENOMENON OF RETRIEVAL OF ABNORMAL PARTICLE CLOUDS TO OUT OF INFECTED ELECTRONS NUCLEONS UNDER DEGENERATIVE A. S. I. F.P. Mol. C.I.C. AND REPLACING ABNORMAL PARTICLE CLOUDS SITE, THROUGH IMPLANTING OR COMBINING OF NORMAL S. Y. E. T. – F.P. – I. I. – P. cl. WITH CNS INTERNAL ELECTRONS NUCLEON PARTICLE COMPOUNDS UNDER REGENERATIVE A.S. I. F.P. Mol. C.I.C., WHICH THIS IS REGENERATIVE A. S. I. F.P. Mol. C.I.C.

THE ABOVE PHENOMENON IS ELIMINATIONS AND RETRIEVAL OF ABONORMAL S. Y. E. T. – F.P. – I. I. – P. cl. TO OUT OF CNS. ELECTRONS NUCLEONS PARTICLE CLOUDS COMPOUNDS. AND IN THEIR PLACE IMPLANTATIONS OR STORAGE, AND RECONSTRUCTIONS OF THE CNS INTERNAL ELECTRON NUCLEON PARTICLE COMPOUNDS WITH NORMAL S. Y. E. T. – F.P. – I. I. – P.cl COMPOUNDS, WHICH THIS IS STORAGE PHENOMENON OF THE CNS ELECTRONS NUCLEONS WITH NORMAL PARTICLE CLOUDS AGAIN. UNDER REGENERATIVE A. S. I. F.P. Mol. C.I.C. (THEREAFTER THE RETRIEVAL OF ABNORMAL PARTICLE CLOUDS TO OUT OF ATOMS). UNDER THESE PROCEDURES THE CNS ELECTRONS NUCLEONS WILL RECONSTRUCT AGAIN WITH NORMAL

PARTICLE CLOUDS, THE PATIENTS WILL REGAIN NORMACY.

THIS PHENOMENON OF ELIMINATION OF ABNORMAL PARTICLE CLOUDS AND REMOVAL OF ABNORMAL S. Y. – F.P. – I.I. – P. cl. TO OUT OF CNS – INTERNAL ELECTRON NUCLEON CONSTRUCTIONS AND REPLACING THEIR PLACE WITH NORMAL S. Y. –F.P. – I.I. – P. cl. COMPOUND CONSTRUCTIONS AGAIN, THIS IS TREATMENT OF CHOICE AND EFFECTIVE ELIMINATIONS OF THE PARTICLE CLOUD INFECTIONS TO OUT OF ELECTRONS AND NUCLEONS. THESE DISEASES AND THEIR TREATMENTS MOSTLY EXPLAINED IN NEXT VOLUME OF THE AUTORS BOOKS. EZZAT E. MAJD POUR, M.D.

FOLLOWING ARE SOME OF MY DISCOVERIES IN THIS VOLUME:

1 - I DISCOVERED THE CAUSE OF PSYCHIATRIC DISORDERS, WHICH ARE PARTICLE CLOUD TRANSMITTED DISEASES, FROM ABNORMAL PARTICLE INFECTED CNS ELECTRONS NUCLEONS TO NORMAL.

2 – I DISCOVERED PLANT'S CENTRAL INTELLIGENCE SYSTEMS, PLANT'S PERIPHERAL INTELLIGENCE SYS.

3 – I DISCOVERED HOW BIOHOSTILE PARTICLE SHOTS DESTROY INTERNAL ATOM PARTICLE COMPOUND CONSTRUCTIONS, ALL ARE WEAPONS OF MASS DESTRUCTIONS, I APPEAL BANNING OF THESE WEAPONS.

4 – I DISCOVERED, IN EMBRYO THE CELL'S MEDIA CHEMICAL FORMULARY CHANGES, CAUSE STEM CELL MUTATIONS IN FETUS, AND THE STEM CELL MUTATION IN EMBRYO PRODUCE CELL'S MEDIA CHEMICAL FORMULARY TO CHANGE, THESE ALTERNATING MUTATIONS IS THE CAUSE FOR STEM CELL DIFFERENTIATIONS, IN EMBRYO THIS PHENOMENON TAKE PLACE SEQUENTIALLY, AUTONOMOUSLY, WITH EXPONENTIAL DEVISION SPEEDS, CAUSE DIFFERENT STEM CELLS MUTATE FROM ONE TO OTHER, PRODUCE DIFFERENT TISSUES, ORGANS, SYSTEMS UNDER EVOLUTION ORDERS, AND CREATE A NEW ANIMAL.

THIS IS FETUS - GENESIS, ANIMAL - GENESIS, AND PLANT –GENESIS PHENOMENON OCCUR FOR ALL LIVING THINGS

UNDER ABOVE ORDERS AND CREATE NEW BORNS UNDER THE EXACT PATHS AND FOOT STEPS OF MOLECULAR EVOLUTION ORDERS.

5 – I DISCOVERED THAT, THE ATOM'S TURN-OVER PROCESS, IS EQUAL TO THE ATOM – GENESIS UNDER THE MOLECULAR EVOLUTION ORDERS AND PATHS. ALSO I DISCOVERED THE EMBRYONIC FETUS – GENESIS IS EQUAL TO THE ANIMAL GENESIS AND PLANT –GENESIS, WHICH OCCURRED UNDER THE EVOLUTION ORDERS AND PATHS, DURING THE WORLDS PAST EVOLUTION HISTORY. THESE RULES CAN BE USED IN CHEMICAL LAB.S, FOR NEW BODY ORGAN -CULTURES (NEW ORGAN –GENESIS), TO BE USED FOR BODY IMPLANTS.

6 – I DISCOVERED FUNDAMENTAL PARTICLES GENERAL CHEMISTRY, ORGANIC CHEMISTRY, PARTICLE'S BIOLOGICAL CHEMISTRY THROUGH A. S. I. F.P. Mol. C.I.C., AND IN CONTINUATIONS LATER UNDER EVOLUTION ORDERS, THE BIO- MOLECULAR AND CELL – BASE AUTONOMOUS SEQUENTIAL CHEMICAL INTERACTION CYCLES CONSTRUCTED PLANTS AND ANIMALS DIFFERENT SPECIES.

7 – I DISCOVERED UNIT OF GRAVITON, UNIT OF MASS, UNIT OF LIGHT ENERGY, UNIT OF ELECTRIC ENERGY, UNIT OF THERMAL ENERGY, ALSO I DISCOVERED PARTICLE CONSTRUCTIONS OF DIFFERENT NANO-UNITS.

8 – I DISCOVERED SONIC –LIGHT FUNDAMENTAL PARTICLES INFORMATION IMAGE PARTICLE- CLOUDS (S. Y. –F.P. – I.I.- P. cl.), HOW PARTICLE CLOUDS CIRCULATE BETWEEN

DIFFERENT ANIMALS AIR-BORN IN CLOSED CIRCUITS, AND DIFFERENT ANIMAL'S CNS ELECTRONS NUCLEONS THROUGH EXOGENOUS PARTICLE CLOUD CIRCULATION SYSTEMS (EX.-PCS) EXCHANGE CNS –PRODUCED S. Y. – F.P. – I.I. –P. cl. WITH EACH OTHER, COMMUNICATE THROUGH EX.-PCS CURRENTS CLOSED CIRCUITS EFFECTIVELY. THESE PCS AND EXCHANGES OF S.Y. –F.P. –I.I. – P.cl. BETWEEEN EACH OTHER IS BASIS OF LEARNING, AND IS BUILDING BLOCKS OF TEACHING, SPEAKING, EDUCATION, LISTENING, TRAINING, PSYCHE-GENESIS, THOUGHT CURRENT –GENESIS, INTELLECTUAL AND KNOWLEDGE ACCUMULATIONS INSIDE CNS. ALL TRANSIT THROUGH EX. PCS.

9 – I DISCOVERED PARTICLE CIRCULATION SYSTEMS INSIDE ELECTRONS, NUCLEONS, NUCLEUS, AND PERIPHERON, ALSO IN PERIPHERAL ATOM SPACES. ATOMS COMMUNICATE THROUGH THESE PCS.

10 – I DISCOVERED LIGHT PARTICLE CONSTRUCTED ATOMS, ELECTRIC PARTICLE CONSTRUCTED ATOMS, THERMAL PARTICLE CONSTRUCTED ATOMS. PARTICLE COLONIES, NANO-UNIT CONSTRUCTIONS.

11 - I DISCOVERED ENERGY PROVIDERS FOR QUANTUM LOCATIONS, WHICH THOSE NANO-ENERGIES ARE USED TO PERFORM NANO-TASKS AT NANO-LOCATIONS.

EZZAT E. MAJD POUR, M.D.

www.ingramcontent.com/pod-product-compliance
Lightning Source LLC
Chambersburg PA
CBHW050206230526
45470CB00001B/257